How To Solve Your Refrigeration and Air Conditioning Service Problems

Carl Geist

Business News Publishing Company
Troy, Michigan
USA

How To Solve Your Refrigeration and Air Conditioning Service Problems

Carl Geist

Copyright© 1986

by

Business News Publishing Company
Troy, Michigan

Library of Congress Cataloging in Publication Data

Geist, Carl.
    How to solve your refrigeration and air conditioning service problems.

    Rev. ed. of: New ways to solve your refrigeration and air conditioning service problems. 1972.
    Includes index.
    1. Refrigeration and refrigerating machinery-Maintenance and repair.
I. Geist, Carl. New ways to solve your refrigeration and air conditioning
  service problems.
II. Title.
TP492.7.G37 1986   621.5'6   85-26993
ISBN 0-912524-26-X

Printed in the United States of America

Contents

Chapter		Page
1	The Serviceman	1
2	Working with Energy	13
3	Superheat and Subcooling	19
4	Head Pressure	35
5	Psychrometrics	41
6	Refrigerants and Controls	53
7	Water and Corrosion	67
8	Air Circulation	93
9	Caring for Condensers	115
10	Using Instruments	133
11	Circuits and Controls	147
12	Evaucation and Driers	187
13	Soldering, Brazing and Leaks	195
14	Lubrication	208
15	Bearings	217
16	Commercial Refrigeration	230
	Afterword	249
	Index	250

Service tools depicted on cover photo courtesy of Robinair

1

The Serviceman

Murphy's Law states: "Anything that can go wrong, will," and, when something does go wrong with an air conditioning, heating or refrigeration system, the call goes out for the **serviceman.**

The **serviceman** is a skilled technician. He knows and understands the fundamental laws of science under which these systems operate. He has the ability to think in a logical manner about the cause and the effect of these laws on the ailing system. He has the instruments necessary to aid him in making the proper diagnosis of the trouble and the tools needed to make the repairs.

Three things, then, go into the makeup of this skilled technician known as the serviceman: knowledge, equipment, and the ability to **think.** The most important of the three is the ability to think. Without this, all the books and tools you can pack into the service truck are worthless. You may change parts until you happen to hit the right one, but you will only be a parts changer and not a serviceman. Sometimes you can change parts for years and never get the job done. For example:

I was sent out to a western state to check a system that had been in operation for seven years. Before leaving I was told that this had been a real trouble job. It had never cooled the building properly, the compressors were unable to run for a summer season without breakdowns, and none of the men who had worked on it had been able to come up with an answer.

I arrived on the job on a Tuesday morning and spent two days checking the installation to get a clear picture of what was installed, and how. Two 60-hp, six-cylinder compressors were in the basement. The shell and tube condensers were on the roof next to the cooling tower and the pumps were beside the condensers. The built-up air-handling unit was in a penthouse. It was a hot deck/cold deck design with 100 percent fresh air available. The pneumatic control system automatically switched to fresh air whenever the outside temperature dropped below 75 degrees. The two cycles were separate and the cooling coils each had three solenoids and expansion valves.

I was familiar with the basic design of this type of system and the installation data book and all blueprints were on the job. Air volumes, coil sizes, and condensing unit capacities seemed to be very well matched. On the basis of the equipment installed, it would seem that the installation should handle the cooling load easily.

My next step was to tear down both compressors to see what parts would be needed. Shafts and bearings were okay, but there had been a lot of valve breakage and it meant new rods, pistons, valve plates and gaskets. The unloader operators were tested with air pressure and none of them would work. When I disassembled them, I found that the internal pistons had been removed. Since these pistons lifted the pins from the suction valve discs to load the machines, it meant only one thing: only three pistons on each machine were pumping. Well, that was my first clue.

I could not find the parts for the unloaders anywhere in the room. Talking to the building maintenance man, I found that, after the factory man worked on it last year, it ran a lot quieter and they did not have as much head pressure trouble as in past years. Beginning to point to the condensers as the trouble, I thought.

The building man and I went up to the roof and he showed me how he removed the rear heads and swabbed the tubes every fall after shutdown. They had had a freeze-up the first winter and two tubes were plugged on one condenser and three on the other. The tubes were fairly clean, but it seemed that the tubes in the lower half were cleaner than in the upper half.

I walked around to the piped end of the condensers and they had the standard three-opening, dual-purpose head. For city water use, you pipe into the bottom, out the top, and plug the middle opening. For tower use, you pump into the top and bottom openings with supply and use the center opening to return the water to the tower. I was eyeballing the pipe from the pump discharge to the condenser head, and it slowly dawned on me that something was wrong. The pump discharge lines entered the bottom and center openings. The top opening was piped to return the water to the tower. Both condensers were piped the same.

Since the only positive circulation would be through the top half of the condensers, see Fig. 1-1, this meant that condenser capacity was half of the compressor capacity. The reason for seven years of trouble seemed to be clear to me. After the condenser piping was changed and the compressors overhauled with capacity control reinstalled, the system proved to be more than adequate to cool the building to everyone's satisfaction. This happened more than ten years ago and the system is still operating and doing a good job.

How did it happen that I had found the trouble after a lot of other men had worked on it for over seven years? For one thing, I knew about dual-purpose condensers. When you are using city water for condensing purposes, you pipe it into the bottom and out the top because the hot gas goes in the top of the condenser and the liquid comes out the bottom. You use the counterflow principle so that the coolest water contacts the coolest gas and the warmest water contacts the hottest gas. The greater the TD (temperature difference) between the two, the more heat transfer you will get. You use all the passes back and forth in the condenser because city water usually is available at about 60 degrees and you can heat it up to about 95 degrees before it goes to the drain.

With increased use of air conditioning and the increased demand for condensing water came the need for water conservation, and cooling towers that recirculated condensing water came into general use.

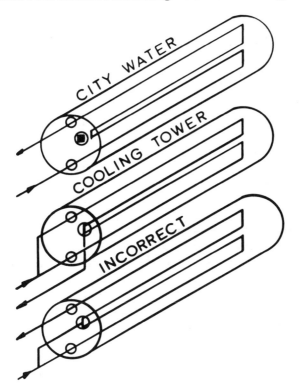

FIG. 1-1   *Right and wrong ways to pipe a dual-purpose condenser.*

Now cooling tower water is not as cool as city water. The normal design of towers enables them to supply water to the condensers at about 85 degrees. You cannot increase the leaving condenser water temperature without raising the head pressure and this will reduce the compressor's capacity. The remedy is to circulate twice as much water through the condenser so that you can pick up the same amount of heat.

How do you put twice as much water through the same condenser? Put half of it into the bottom, as before, and take it out the center after it has picked up its normal amount of heat. Put the other half into the top and take it out the center after it has picked up its heat rise. You can use the same condenser, with the only necessary change being the piping to the center opening in the head.

Practically all manufacturers use this design. Regardless of whether the customer is going to use city water, well water, or a cooling tower, this one condenser will do the job. Knowing this, and thinking about how the condenser should be piped as I looked over the installation, it seemed impossible for me not to spot the trouble.

The size of the installation has nothing to do with the requirement that the serviceman needs to think on every service job. Let's take a look at a residential service call. It was a central air conditioning system that consisted of a 4-hp, air-cooled condensing unit and a gas-fired furnace with an *A* coil in the discharge plenum. The complaint was that it would cool for a while and then quit, run but not cool.

I checked the filters and they were clean. I checked the fan, belt, and pulleys and they were okay. The sheet metal plenum did not have any access openings so I could not check the coil for dirt on the fins. The sight glass in the liquid line was clear so I had plenty of R-22. I found gauge valves on the condensing unit so I attached my suction and discharge gauges. The air entering the condenser was 86 degrees and the discharge pressure was 170 psi. That meant I had a 90 degree condensing temperature and an efficient condenser. Suction pressure was 52 psi. The corresponding 28 degree evaporator temperature was too low. I had to see if the coils were dirty. I found a small rectangle of sheet aluminum in the truck and cut an access hole in the plenum that was one inch smaller than my aluminum panel. The coils were clean so I fastened the aluminum cover over the hole and sealed it with duct tape.

I was now convinced that I was short of air across the evaporator coil for some reason. I went out to the truck and got my three-lead resistance thermometer, the clip-on ammeter, and the manometer, Fig. 1-2. The air was entering the coil at 74 degrees and leaving at 40 degrees. The static pressure across the fan and furnace heat exchanger was 0.4 in. water gauge. The fan motor was drawing 4.4 amperes. I took the motor out of the bracket to see the nameplate and found it was ¼ hp, rated at 4.2 amperes.

Consider the facts. A 34 degree drop across the coil is too much for a standard air conditioning system. A 28 degree coil temperature means that condensation on the coil is freezing and will soon accumulate to block air flow. A ¼-hp motor is all right for a heating furnace fan but is not big enough for a 4-ton system that could require from 1200 to 1600 cfm of air. The motor was running at slightly over its rated capacity and I could not get any more work out of it.

The only answer was to install a larger motor and speed up the fan. The lady of the house got her husband on the phone. I told him what I had found, and explained how more air over the coil would stop it from freezing up and killing the cooling. He said he had bought an air conditioned home and he wanted the air conditioning to work. If it needed a larger fan motor, get and put it on. Well, I made a quick trip to the wholesaler and picked up a 1/3-hp motor, an adjustable sheave that was ½ in. larger and a belt that was 1 in. longer. With the new fan motor installed, I adjusted the sheave to load

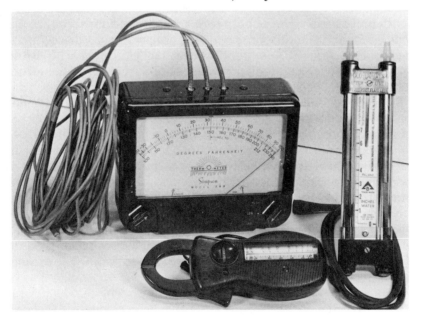

FIG. 1-2   Service tools: three-lead resistance thermometer, manometer, clip-on ammeter.

the motor to its rated 6.2 amperes. Static pressure was now 0.6 in. Air was leaving the coil at 52 degrees with 74 degree entering air. Suction pressure was 57 psi. The homeowner tried the system for a week before he paid the bill, but he was very happy with the results. His paying the bill made me happy too.

What happened to set up this service call? Did someone install the unit and fail to check the fan capacity or try to get by with the present motor on the furnace? Could be. Did other servicemen make calls on this job and fail to find the trouble? Yes! The owner had commented on this fact. Why did they fail to find the trouble? Because, when they arrived and turned on the system, the coil had defrosted and it would do a good job for a while. They could feel the cold air discharging from the registers and assume it was some momentary problem that had corrected itself by the time they arrived. Did they install service gauges and check operating pressures? I have no way of knowing what they did or if they even noted the low suction pressure.

What made it possible for me to correct the problem and make the customer happy? I knew that it had happened before because the customer told me so. The 28 degree coil temperature was below freezing and was my first clue to a possible solution. I had enough instruments and I used them to help find an answer. Knowledge, some from books and some from experience, was one tool I used. Instruments, properly used were another tool. Logical thinking, based on general knowledge and the knowledge gained from the instruments, made it possible to find the answer to the problem.

Let's take a look at another service call on a residential air conditioning system. I found the unit off on reset. The circuit breaker had not tripped so I pushed the control box reset button and the condensing unit started. I was watching it through the back panel and trying to figure out how to check for the cause of the trouble. There were no gauge connections and the electrical control box was hard to get to. I noticed that the two condenser fans were slow to get up to speed and were very noisy. I stopped the unit and checked the fans. The left fan had four blades with a shallow pitch. The right fan had three blades with a very deep pitch. I could see that the condenser fins behind the left blade had been flattened, as if the fan had scraped it at some time. When I checked with the lady of the house, she said that a serviceman had installed a new fan last year.

I removed the entire housing from the condensing unit so that I could get at the fins on the condenser. I found that the left fan was loose on the motor shaft. The right fan was loose on the fan hub. The left fan was fastened to the hub with screws and the right fan was held to the hub by a riveted washer and two rubber gaskets. I straightened the fins with a fin comb and got two new fans from the distributor of the condensing unit. That way I could be reasonably sure that I was moving the right amount of air across the condenser coil.

The serviceman who replaced the fan the year before certainly should have straightened the fins before installing a new fan. Anything that blocks air flow cuts down on capacity. Replacing a high pitch blade with a low pitch blade certainly cuts down on the air flow, and this cuts down unit capacity. The unit had cut out on head pressure control for a while and finally locked out on motor overload when they called me. Knowledge and a little thinking on the part of the previous serviceman would have prevented the second call.

A good many things go together to make up an efficient air conditioning coil, fan, and condensing unit operation. Knowledge, instruments, and thinking go together to make a good service call that will make the customer happy. The serviceman who does a thorough job of checking out a system will find that he has very few call-backs.

I will never forget one service call I had where I thought I had done a very good job but still got a call-back. Cooling in this store had gotten so bad that the customer had two 25-ton variable capacity compressors removed and a 50-ton, belt-driven compressor installed in their place. This old two-cylinder upright was rocking along with a frosted suction line and crankcase when I was sent out for my first visit to the job. I had been told what was happening before I left the shop and commented that, if the compressor was icing up, we either had a dirty coil or insufficient air was causing the icing. I was told that the coils had been cleaned and the fan motors checked and found okay. It was up to me to find the answer.

I found that the coils were clean on the surface but were loaded with oily dirt deep between the fins. The air was being moved by six squirrel-cage fans, three to a shaft with two motors. The curved inner surface of the fan blades contained sufficient oily dirt to effectively cut down the capacity to move air. Dirt was the only problem: dirt that cut down on air delivered, air that contacted the coil surface; dirt that covered the prime and secondary coil surfaces and cut down heat transmission. My knowledge of the fundamental laws covering the operation of air conditioning systems told me that this dirt was the only problem.

It took me two days to properly clean the fans and the cooling coil. After I was through, the increased air movement through the store was very noticeable. The suction pressure was higher and the frost gone from the compressor. The improved wet and dry-bulb temperatures of the air leaving the coil made me happy enough to say to the store manager, "Your troubles are over."

The next morning I was sent back to the store as soon as I arrived at the shop. My instructions were, "Get out there quick, they've got trouble." They did have trouble. The store temperature was 62 degrees. All the zone thermostats had been set down as far as possible when they were having trouble with the cooling. When I picked up after getting all the cleaning done, I had failed to make a check of the thermostat settings. It just goes to show that I had quit thinking too soon on this service call.

SERVICE PROCEDURES

The serviceman who knows his fundamentals and thinks can come up with the answer to almost any service problem that he encounters. But thinking should not be confined to problem solving. Thinking should be part of all service work. If it is not a part of all service work, then we run into the lazy, sloppy serviceman.

I always try to do a neat, workmanlike job on service and it bothers me to run into troubles caused by lazy servicemen and sledgehammer mechanics. Take the call where you have to put service gauges on the compressor. You take the valve cap off, loosen the packing gland nut, back seat the valve, remove the gauge port plug, insert an adaptor fitting and attach the gauge. It sounds like a simple procedure and it is, but a lazy, sloppy serviceman can surely foul it up. The average machined-brass valve cap has hex flats and can easily be removed with an adjustable wrench. A lazy man, who does not take the time to adjust the wrench for a tight fit, will round off the corners of the hex and ruin the cap. Then he takes a pipe wrench or a pair of pliers and proceeds to chew off some more metal. After he gets the cap off, he loses the copper gasket and sometimes the whole cap.

Unless the packing is loose enough to blow his hand away, I don't think the Sloppy Joes will even bother to touch the packing gland nut. If he does, you can be sure he will use pliers or a loose wrench and round off or burr the gland. Most gland nuts have only two flats on them. Once you ruin them, the next man is in trouble. Then, Sloppy Joe can't be bothered to pick up a valve wrench. He will either round the square end, so the next man cannot get a valve wrench on it, or he will burr the stem so that when the valve is front-seated the burrs on the stem chew out the packing. If it is a steel stem and if it is the suction valve, and if it is not protected by the valve cap that was thrown away sometime back, the stem will be rusty and corroded. Does Sloppy Joe clean the stem before he front seats the valve? He does not! Run her in and chew up the packing. Can't take time to think, gotta get her fixed.

You would think by now that Sloppy Joe had things pretty well loused up, but this is only the beginning. Next, he slaps that big adjustable wrench on that little old plug in the gauge port and proceeds to round off all six points of the hex. It is still not quite round, so he uses a pair of pliers to take off .01963 in. of metal between the remaining flats. By this time he has barked a knuckle or two on the burrs of the valve stem and, after stating what he thinks of the so-and-so who put the plug in that tight, he gets a pipe wrench on the plug and gets it out.

Now, he may throw the plug away at this point and leave the port open to gather dirt and foul up the works after he is gone. I generally find that he puts it back with iron cement and his pipe wrench and says, "They'll never get that out again." Be that as it may, after he gets the plug out, he rummages around in the bottom of his box until he finds a pipe-to-flare adaptor to screw into the port. He knows it is an adaptor, but you and I would never recognize it under the grease and dirt it has collected.

Is the inside of the adaptor dirty? We'll fix that. Just crack the valve, blow the dirt out. Well, waddayaknow, must be real low on gas if it is pulling a vacuum. Say, hold your finger on that adaptor while I fish my pliers out from under the base will you? Don't want too much air to get in the system. Valve stem sure turns hard. Packing must be dry. Okay, that did it.

That is just a sample of what Sloppy Joe can do. We haven't gotten to the discharge valve yet, nor the adjustable superheat expansion valves, nor the manual opening stems on solenoid valves. One of the favorite tricks of the sledgehammer mechanic is often found on the larger compressors with two, four or six-inch lines. These valves usually have cast iron caps and require a pipe wrench or one of the new hex jaw wrenches to take them off. Rather than go back to the truck for a big wrench, he gets out his hammer and breaks the cap loose with glancing hammer blows. This works for a few times, but there comes a day when the cast iron breaks. I have found some of these valve caps wrapped with friction tape to hold the pieces together. Last spring I found the suction and discharge valve caps on two 75-ton compressors battered enough to be slightly out of round. They had been put back on with refrigerant pipe dope and I had to use a 36 with a *cheater* to get them off. Cleaning off the dope with acetone, I found the packing glands had also been doped. I finally ended up ordering new packing, packing glands and caps. It cost the customer over $125.00 just to get these four valves working again.

Several years ago I was asked to check six 20 and 30-ton package units in a department store for leaks, because they had found they were buying a lot of refrigerant each year. I found a total of 21 leaks on the six units, and every one of them was due to leaking valve stems or sight glasses. I sold them valve caps where they were missing, sight glass gasket and glass kits, and caps where needed, cleaned all the valve stems with 500 grit tape, lubricated all the packing and stems with oil and replaced practically all of the gauge port plugs. It was a neat, serviceable job when I got through and they are not adding refrigerant anymore. I know, because they are still my customers. Two years ago I sold them on changing the oil and installing a replaceable core drier on each unit.

I always take the time to get the right wrench from the truck when it is needed. After all, I am being paid by the hour, not by the job. Most of the 1/8-in. and 1/4-in. gauge port plugs can be handled with a 1/2 x 9/16 box wrench, and 7/16, 5/8 and 11/16 are not uncommon. The average 5-piece box wrench set will handle all of them. Good 8, 10 and 12-in. adjustable wrenches are fine for valve caps. I mean good ones though. The worn out knucklebuster you find in Sloppy Joe's tool box is costing him money. I have a 14-in. aluminum pipe wrench that never was a very good pipe wrench. I ground the teeth off and it makes a good valve cap wrench for caps too big for the 12-in. adjustable.

On the small hermetic condensing units with valved gauge connections, I find that most of these valves are loose where they bolt on to the frame. Too many of the boys are using just one wrench to take the cap off. If they would use another wrench on the valve body as a backup, it would stop a lot of damage. I carry a small stock of ⅛-in. and ¼-in. hex head pipe plugs and ¼-in. and ⅜-in. flare caps. When I find these things damaged on the unit, I show them to the owner and try to sell him new ones. They usually buy them, too. They are not expensive, but they carry a good markup for me.

I've often wondered why the Sloppy Joes are that way. It must be that they just don't think. If they ever did start thinking, they would realize that they are hurting themselves as much as the customers. The customer can see sloppy work and all too often this is how he judges the man.

Speaking about sloppy work, one of the easiest ways I know of to foul up a job and lose customers is Greasy Gus' method. Greasy Gus is a lubrication specialist. When he leaves the job every bearing has been oiled and greased. The customer can see the oil dripping from the motors and the grease that has been forced out around the shaft. What the customer can't see is the oil running into the motor windings to collect dust and create a fire hazard. He can see the grease around the fitting on the ball bearing motor but he does not see the bottom plug that should have been removed before lubricating the bearing. He can't see the ruptured seal inside the motor and the grease that is being thrown into the field windings.

I remember a 7½-hp cooling tower fan motor with a 36 in. cast aluminum fan on it that was overloading and running hot. We could not get the fan off so we pulled the motor and fan, and took them to the shop so we could press the fan off the shaft. When we took the end bell off, we found the motor was packed with grease to the point where all we could see was the shaft sticking up out of the grease. The bearings had double plugs in them but the relief plugs had never been taken out. Normally, you would take out the relief plug and grease the bearing once a year. My guess is that the motor had been greased once a month without touching the relief plugs. The motor manufacturers are familiar with this trouble. Some of them are installing a spring cover oil cap to act as a relief plug. The spring cover will prevent pressure buildup that could rupture the grease seal.

One of the worst results of oiling a motor or fan bearing too much is when it is above the cooling coil in a package unit. Two bad things can happen. The oil dripping on the cooling coil coats the fins and this collects dust. Before long the coil is dirty enough to cut down the heat transfer and capacity is lost. This type of coil fouling usually occurs deep in the coil and cannot be brushed off. It requires solvents or detergents and water or steam pressure to get the coil clean all the way through. The other thing that happens is that oil from the fan bearings flows along the shaft and is thrown onto the blades of the squirrel-cage fan. The centrifugal force spreads this oil evenly over the blades and a gradual dirt buildup in the oil can fill the inner curve of the blades. I have measured a loss of almost 30 percent in air volume due to dirty fans. What amazes me is that so many men can look right at those dirty fans and not realize that they are seeing the reason why the customer is complaining about insufficient cooling. That is why we have the Sloppy Joes and the Greasy Guses. They just don't think.

Thinking is important, but knowledge of the fundamental laws that govern the operation of air conditioning, heating and refrigeration systems is necessary so that the serviceman can think logically about cause and effect when diagnosing troubles. Let's take a look at these laws in the next chapter.

2

Working with Energy

When you are working on air conditioning, heating or refrigeration you are dealing with energy. If we are going to think logically about energy, we have to have a place to start. So let's start at the beginning.

All matter is made up of atoms. Atoms are made up of positive particles called protons, negative particles called electrons, and neutral particles called neutrons: positive, negative, neutral. So we are in the field of electricity from the very beginning.

The simplest atom is the hydrogen atom: one proton with one electron in orbital motion around it. Hydrogen is an element. So is oxygen, carbon, calcium, iron, magnesium. Elements differ in their atomic structure. That is, the atoms of these elements have different numbers of protons, electrons, and neutrons in their make-up.

We are not concerned with the make-up of these atoms but we must accept the fact that the atoms of these elements have electrons in motion around the proton. As long as these electrons are in motion, the atom contains energy in the form of heat. To stop this heat in the atom you have to cool it down to absolute zero. So far, man has not been able to accomplish this. He has come within a few hundredths of a degree of it, but no closer.

Atoms can combine with other atoms to form molecules. Combine atoms of sodium with atoms of chlorine and you get molecules of sodium chloride or common table salt. Combine atoms of chlorine, fluorine, carbon and you get molecules of dichlorodifluoromethane or what we call Refrigerant-12. We are not concerned with the how of these compounds, but we must accept the fact that all the atoms that go to make up compounds have electrons in motion in them and because of this fact, these compounds contain energy in the form of heat.

This brings up the question of what is cold. Hot and cold are relative terms. We say something is hot or cold in comparison to something else. **Think** about it: now here's a *hot* item, that's a *hot* number, I got the *cold* shoulder. To the serviceman, heat is a form of energy, which brings us to the principal physical law we work with: the Second Law of Thermodynamics.

Take two or more substances with different amounts of heat energy and place them so that they can exchange this energy. The substance with the most energy will transfer energy to the substance with less energy until both are at the same energy level. In simple language, place a hot brick on a cold brick and after a while both bricks will be at the same temperature, somewhere in between the two original temperatures.

Place a sheet of insulation between the bricks and it will take a lot longer for the temperatures to equalize. Insulation is made of substances that are slow to conduct energy and have lots of dead air spaces. You haven't stopped the energy transfer, but you have made it travel a lot farther and gained time by slowing down this transfer. Put the hot substance in a vacuum bottle and it will take even longer for it to lose its heat energy. You have surrounded it with a near vacuum in which the air molecules are far apart. In addition, the outer wall is covered with a reflective surface to bounce the energy back into the hot substance. It is not perfect but you have gained a lot of time in the race to stop the heat energy loss. Noticed how long coffee will stay hot in your vacuum bottle and how quickly it cools in your cup?

This Second Law of Thermodynamics is the basic law that governs all of our actions in refrigeration, heating, and air conditioning. We put a hot fire inside a chamber made of a metal that readily conducts heat and pass air over the other side of this chamber so that the heat inside will give up its energy to the air outside. We use this heated air to warm our homes and buildings. Energy exchange takes place because of the Second Law of Thermodynamics.

In refrigeration and air conditioning we create a condition where we take heat away from air or other substances and, because of this Second Law of Thermodynamics, we are able to refrigerate, freeze, and cool as necessary. We also humidify and dehumidify because of this law, but we will get into the how of this later.

Boiled down, the Second Law of Thermodynamics means that all energy eventually reaches a state of equilibrium. Volumes, pressures, and temperatures interchange until they reach a stable energy state. Only by upsetting this state of equilibrium can we make a system pump heat.

In the first chapter I told about a service call where the compressor crankcase was icing up and I blamed this on either dirty coils or insufficient air over the coils. As it turned out, I had a combination of both these things. I based my reasoning on the fact that the refrigerant was coming back to the compressor too cold. It had not absorbed enough heat energy from the air passing over the coil to bring the refrigerant up to the designed temperature. Something was slowing down the transfer of heat energy and it had to be a form of insulation between the two substances, air and refrigerant gas. Since the only thing that was supposed to separate these two was a highly conductive copper coil with aluminum fins, the logical answer was dirt collecting on the surface of the metal. When I found the dirt and cleaned the coil surfaces, I got a good exchange of heat energy and the system performed as designed.

One thing that can make me *hot under the collar* (there's that relative term again) is to have a service call with low back pressure and some icing and ask a serviceman what he thinks can be causing it. It is surprising how often they will say the compressor might be too big or the coil might be too small or the expansion valve may be bad. If this were a brand new installation being started for the first time, it is barely possible that these might be the reasons, but never on a system that has been operating and giving satisfactory results for some time.

Something has to be upsetting the balance of energy exchange between the air over the coil and the refrigerant in the coil. 90 percent of the time the answer is dirt that is insulating the coil and cutting down the air flow. A friend of mine operates a shop where he has an assembly line set up for servicing window air conditioners. The first step, for all units brought in, is to take off panels and steam clean the units with special attention to cleaning the evaporator and condenser coils. My friend says that 90 percent of the units operate satisfactorily after they are cleaned this way and then put on the test station. It's something to think about.

Before we leave the subject of energy, let's look at the three ways that heat can be transferred: radiation, convection and conduction.

Radiation: heat from the sun, from a fireplace, from a heat cable in the ceiling.

Convection: heat absorbed by a fluid such as air or water, causes that fluid to expand and become lighter than the surrounding fluid. It moves up because it is lighter. As it moves up cool fluid moves in to replace it. This sets up a circulation in the fluid and heat is transferred from the source of heat to a place where the warmed fluid can give up its heat to a cooler substance. The old gravity type warm-air furnaces were good examples of this. Present day furnaces are forced convection furnaces because we have added a fan to them. Now we can get more air through smaller ducts and even send it downhill to do the job.

Cooled air sets up convection currents too, only the air flows down instead of up. I can remember service calls where window units had been installed and the occupants complained that the first floor did not get cool but the basement was like ice. The remedy was to cover or close the return air grille and supply registers that let all the cold air down into the basement.

Conduction: Hold your hands out to a radiator and you can feel the radiant heat from it. Blow some smoke around it and you can see the convection currents of warm air rising from it. Touch the surface and you can burn your hands. That's conduction. Enclose a source of cold, or heat, in a container that is conductive and it will transfer heat energy through the container wall as the temperature levels try to equalize.

I did not use the term *a source of cold* to be confusing. When we want to cool something we have to take heat away from it. We have to have a receptacle for this heat. We create this receptacle in a substance by taking heat away from it so that it will have less energy in it and be able to absorb energy from the other substance.

The next law the serviceman must know is the Law of Conservation of Energy: that energy cannot be created nor can it be destroyed. It can only be transformed into some other form: sunlight into trees and plants; plants into coal; coal into electricity; liquid into gas; heat into pressure; pressure into heat; pressure, heat and carbon into diamonds; hydrogen into sunlight and helium. Accept the fact that energy cannot be created or destroyed and we can get into the laws of gases that govern the actions of all refrigeration and air conditioning systems.

Charles' Law

A. At a constant volume, the pressure of a gas varies directly as the absolute temperature.

B. At a constant pressure, the volume of a gas varies directly as the absolute temperature.

Why absolute temperature? Atoms have heat energy as long as they are above absolute zero. Scientists working with energy in this form have to use absolute temperatures in order to get correct answers down to the last decimal point.

What does part A mean to us as servicemen? If we have a cylinder of refrigerant, the pressure of both the liquid refrigerant and the refrigerant gas in that container is going to increase as the temperature increases. If our refrigerant container is the condenser/receiver of an air conditioning system, we can reduce the pressure by passing water through the condenser tubes and taking heat away from the refrigerant. We use water that is at a lower temperature than the refrigerant and heat travels to it because of the Second Law of Thermodynamics. Pressure goes down because of part A of Charles' Law.

To understand part B, we must look at the refrigerant in the evaporator cooling coil. Refrigerant is in two states, liquid and gas. The volume is fixed by the size of the coil. If we pass air that is warmer than the refrigerant over the outside of the cooling coil, heat will be taken from the air into the refrigerant because of the Second Law of Thermodynamics. The liquid will expand and actually boil as it changes to a gas that is trying to increase in volume. The volume is fixed by the size of the coil so the pressure will try to increase in obedience to part A. We are holding the pressure constant by drawing off the increase in gas volume with the suction effect of the compressor.

So far we have discussed the Second Law of Thermodynamics and Charles' Law. What it boils down to is that heat will always travel from high to low temperature, refrigerant pressure will increase if you add heat, and refrigerant volume will increase if you add heat. In the process, we have also thrown in the fact that a refrigerant will change from a liquid to a gas. This change of state depends on the amount of energy in the refrigerant. A cylinder of refrigerant maintains a constant volume because of the refrigerant cylinder walls. At

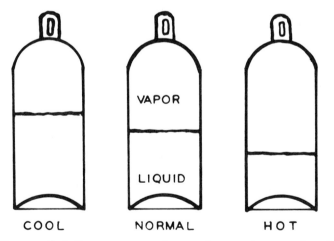

FIG. 2-1 Relative amounts of vapor in a cool, normal and hot container.

a high temperature, we will have high pressure, a large percentage of gas and a small percentage of liquid. At a low temperature, we will have low pressure, a small percentage of gas and a large percentage of liquid, Fig. 2-1.

We have been working with constant volumes and constant pressures with Charles' Law. What happens if we work with constant temperatures? Now we are getting into Boyle's Law.

Boyle's Law

A. The volume of a gas varies inversely as the absolute pressure providing the temperature remains constant.

B. The absolute pressure of a gas varies inversely as the volume providing the temperature remains constant.

We have had refrigerant in the condenser/receiver and in the evaporator coil. We had liquid refrigerant and we changed it to gas under the application of the Second Law of Thermodynamics and Charles' Law. Now let's put that gas in the compressor and apply Boyle's Law. As the piston goes down it creates a vacuum in the cylinder. Since nature abhors a vacuum, refrigerant gas rushes in to fill this space. When the piston comes up, the volume of the gas is decreased by the compressing effect. Following Boyle's Law B, pressure increases as volume decreases. We have already put Boyle's Law A to work in the evaporator with the suction effect of the compressor on it.

What happens to the heat energy that is in this low temperature, low pressure, high volume gas when it is compressed to a low volume, high pressure gas? It becomes high temperature gas. We have not taken any heat away from it so the same heat in a smaller space results in higher temperatures. In fact, we have added heat to it in the form of friction heat from the compressing machinery. Now we will send this gas to the condenser/receiver where we know what will happen because of the laws of gases. It will give up its heat to the condensing medium and change to a mixture of liquid and gas. We will send the liquid to the evaporator coil and reduce the pressure on it. It will pick up heat and change back to a gas in the process that obeys the laws of gases. To the compressor, then to the condenser, then back to the coil, in an endless cycle that pumps heat for us in obedience to these laws.

3

Superheat and Subcooling

The drawing of the three cylinders of refrigerant, Fig. 2-1, shows refrigerant in a saturated state. It is a closed cylinder. Pressure is constant. Refrigerant is present in both liquid and vapor states. The pressure of the vapor balances the pressure of the liquid. Energy is in balance. Change the pressure and the ratio of liquid to gas will change.

If heat is added to increase the pressure, some of the liquid will change to vapor. If heat is taken away to reduce the pressure, some of the vapor will change to liquid. What this means to the serviceman is, that for every definite pressure, there is a definite temperature for that particular refrigerant. This is the saturated temperature. Add heat at this pressure and you are going to evaporate refrigerant. Take away heat and you are going to condense refrigerant.

A refrigeration system operates at a more or less constant pressure in the high side and the low side. We add heat to the low side and we take heat away from the high side. The serviceman who uses the saturated pressure/temperature tables and understands the operation of the system can see in his mind just what is happening inside these parts of the system. When he uses his pressure gauges and thermometers to read pressures and temperatures inside the system, he compares the readings to the tables for these readings. The variations from saturated pressure/temperature tell him what is actually happening inside the system. Using this knowledge, the serviceman can reason out what adjustments, corrections or repairs are necessary.

Let's take a look at the saturated pressure/temperature table for water, Table 3-1. Water is the refrigerant used in lithium bromide absorption water chillers of some air conditioning systems. Water is also the media used in steam heating systems. The serviceman working on absorption systems and the serviceman working on steam heating systems use this same table in their work. There are four pressure tables and two temperature tables. The dominant column in this table is degrees Fahrenheit. That is the system of temperature measurement commonly used in the U.S. at this time.

Pressures below atmospheric are usually measured in inches of mercury absolute. Some absorption machine manufacturers measure these pressures in millimeters mercury absolute and when the United States changes to the metric system, this may be replaced

PROPERTIES OF WATER

Gauge	PSIA	in. Hg	mm. Hg	°F	°C	Enthalpy	Latent
		ABSOLUTE				BTU / POUND	
29.74	.09	.18	4.6	32	0	1075	1075
29.71	.10	.20	5.2	35	1.7	1076	1073
29.67	.12	.25	6.3	40	4.4	1077	1069
29.62	.15	.30	7.6	45	7.2	1081	1068
29.56	.18	.36	9.2	50	10.0	1083	1065
29.49	.21	.44	11.1	55	12.8	1085	1059
29.40	.26	.52	13.5	60	15.6	1087	1058
29.28	.30	.62	15.6	65	18.3	1089	1055
29.18	.36	.74	18.6	70	21.1	1092	1052
29.05	.43	.87	22.2	75	23.9	1094	1050
28.89	.51	1.03	25.4	80	26.7	1097	1047
28.71	.60	1.21	30.8	85	29.4	1098	1045
28.50	.70	1.42	36.6	90	32.2	1101	1042
28.26	.82	1.67	42.2	95	35.0	1102	1039
27.99	.95	1.93	49.1	100	37.8	1105	1036
27.68	1.10	2.24	56.6	105	40.6	1107	1034
27.32	1.27	2.60	66.9	110	43.3	1109	1031
26.93	1.44	3.00	76.1	115	46.1	1111	1028
26.47	1.69	3.45	88.2	120	48.9	1113	1025
25.97	1.94	3.96	100.2	125	51.7	1115	1022
25.39	2.22	4.53	115.1	130	54.4	1117	1019
24.75	2.54	5.17	131.2	135	57.2	1119	1017
24.04	2.89	5.88	149.3	140	60.0	1121	1014
22.35	3.72	7.57	191.1	150	65.6	1125	1008
20.27	4.74	9.66	245.8	160	71.1	1130	1002
14.63	7.51	15.29	388.9	180	82.2	1138	990
6.45	11.53	23.47	596.8	200	93.3	1146	978
0	14.70	29.92	759.9	212	100.0	1150	970
5	20	40.83	1036.3	228	108.9	1156	960
15	30	61.25	1555.7	250	121.1	1164	945
30	45	91.87	2336.8	274	134.4	1172	928

in. Hg vacuum (spanning rows 28.71 to 27.32)

Table 3-1

with a different measurement scale.

Reading this scale, you can see that at zero pounds per square inch gauge the corresponding atmospheric pressure is 14.696 pounds per square inch absolute. The corresponding pressure in inches of mercury absolute is 29.92 which should be familiar to you who have heard the barometric pressure given on the weather reports. At this pressure water boils at the familiar 212 degrees Fahrenheit. The enthalpy (total heat above 32 degrees) is 1150 Btu per pound of water. Take away the amount of heat required to raise the water temperature to the boiling point and you have left the amount of heat necessary to change the water to steam (latent heat of vaporization): 970 Btu.

The absorption machine works because the pressure on the water in the evaporator is 6.3 mm Hg absolute. At this pressure water will boil at 40 degrees F if heat is added to it. When the water used as the refrigerant contacts the pipes carrying the chilled water at about 55 degrees F, it absorbs heat and vaporizes. Each pound of water thus vaporized absorbs 1069 Btu's of heat energy.

When the heating serviceman checks the amount of condensate from a steam coil that is supplied with steam at 30 psig, he knows that each pound of water gave up 928 Btu's of heat as it condensed.

The tables for refrigerants are not as elaborate. Pressures are given in psig (pounds per square inch gauge) and inches mercury vacuum. The corresponding temperatures (saturated) are given in degrees Fahrenheit.

When the serviceman is checking a system he will often get readings that are different from the readings shown on the tables. When he gets these readings he knows that the readings were taken at a point in the system where saturated conditions do not prevail. If the temperature read is above the saturated temperature for the prevailing pressure, then we know that the refrigerant has been superheated. Refrigerant cannot be superheated until all the liquid has been vaporized. When we take a temperature reading on a suction line and find that the reading indicates superheat, we know that we have vaporized all the liquid, and the suction line is carrying only refrigerant vapor.

When we take a reading and find that the actual temperature is below the corresponding saturated temperature for that pressure, we know there is no vapor at that point. Refrigerant cannot be subcooled as long as vapor is present. Heat energy taken away from a mixture of gas and liquid will always condense gas to a liquid before it will change the temperature. These are two important factors in service work. Let's take a look at superheat first.

SATURATED PRESSURE/TEMPERATURE TABLES

Fahr.	Cels.	R-11	R-12	R-13	R-22
-160	-105			*in. Hg* 22.7	
-140	-95		29.4	15.3	29.1
-120	-83.6		28.6	4.9	27.7
-100	-72.9		27	7.3	*in. Hg vacuum* 24.8
- 90	-67.3		25.7	14.1	22.7
- 80	-62.3		24	22.3	20
- 70	-56.6		21.9	32.3	16.5
- 60	-51.0		19	42.3	*in. Hg vacuum* 13.5
- 50	-45.5	29	15.4	57.3	6.0
- 48	-44.4	28.9	*in. Hg vacuum* 14.6	60.0	4.7
- 46	-43.3	28.8	13.8	63.0	3.3
- 44	-42.2	28.7	12.9	66.2	1.8
- 42	-41.1	28.6	11.9	69.4	0.3
- 40	-40.0	28.4	*in. Hg vacuum* 11.0	72.7	0.6
- 38	-38.9	28.3	10.0	76.2	1.4
- 36	-37.8	28.2	8.9	79.7	2.3
- 34	-36.7	28.1	7.8	83.3	3.2
- 32	-35.6	28	6.7	87.1	4.1
- 30	-34.4	27.8	5.5	90.9	5.0
- 28	-33.3	*in. Hg vacuum* 27.7	4.3	94.9	6.0
- 26	-32.2	27.5	3.0	98.9	7.0
- 24	-31.1	27.4	1.6	103	8.1
- 22	-30.0	*in. Hg vacuum* 27.2	0.3	107.3	9.2
- 20	-28.9	*in. Hg* 27	0.6	111.7	10.3
- 18	-27.8	26.9	1.3	116.2	11.5
- 16	-26.6	26.7	2.1	120.8	12.7
- 14	-25.3	26.5	2.8	125.7	13.9
- 12	-24.2	26.2	3.7	130.5	15.2
- 10	-23.2	26	4.5	135.4	16.6
- 8	-22.1	25.8	5.4	140.5	18
- 6	-21.1	25.5	6.3	145.7	19.4
- 4	-20.0	25.3	7.2	151.1	20.9
- 2	-18.9	25	8.2	156.5	22.5
0	-17.8	24.7	9.2	162.2	24.1
2	-16.5	24.4	10.2	167.9	25.7

Table 3-2

SATURATED PRESSURE/TEMPERATURE TABLES

Fahr.	R-113	R-114	R-500	R-502	R-717
-100			*26.4*	*23.3*	*27.4*
- 90			*24.9*	*20.7*	*26.1*
- 80			*22.9*	*16.5*	*24.3*
- 70			*20.3*	*12.5* (*in. Hg vacuum*)	*21.9*
- 60			*17*	*7*	*18.6*
- 50		*27.2*	*12.8*	*0*	*14.3*
- 48		*27*	*11.6*	0.8	*13.3*
- 46		*26.8*	*10.9*	1.6	*12.2*
- 44		*26.6*	*9.5* (*in. Hg vacuum*)	2.5	*11.1* (*in. Hg vacuum*)
- 42		*26.3*	*8.5*	3.4	*10*
- 40		*26.1*	*7.7*	4.3	*8.7*
- 38		*25.9*	*6.7*	5.2	*7.4*
- 36		*25.6*	*5.4*	6.2	*6.1*
- 34		*25.3*	*4.2*	7.2	*4.7*
- 32		*25*	*2.8*	8.3	*3.2*
- 30	*29.3*	*24.7*	*1.4*	9.4	*1.6*
- 28	*29.3*	*24.4*	*0*	10.5	*0*
- 26	*29.2*	*24*	0.8	11.7	0.8
- 24	*29.2*	*23.7*	1.5	12.9	1.7
- 22	*29.1*	*23.3* (*in. Hg vacuum*)	2.3	14.2	2.6
- 20	*29.1*	*22.9*	3.1	15.5	3.6
- 18	*29*	*22.5*	4	16.9	4.6
- 16	*28.9*	*22.1*	4.9	18.3	5.6
- 14	*28.9*	*21.6*	5.8	19.7	6.7
- 12	*28.8* (*in. Hg vacuum*)	*21.1*	6.8	20.2	7.9
- 10	*28.7*	*20.6*	7.8	22.8	9
- 8	*28.6*	*20.1*	8.8	24.4	10.3
- 6	*25.5*	*19.6*	9.9	26	11.6
- 4	*28.4*	*19*	11	27.7	12.9
- 2	*28.3*	*18.4*	12.1	29.4	14.3
0	*28.2*	*17.8*	13.3	31.2	15.7
2	*28.1*	*17.2*	14.5	33.1	17.2

Table 3-2

SATURATED PRESSURE/TEMPERATURE TABLES

in. Hg vacuum

Fahr.	Cels.	R-11	R-12	R-13	R-22
4	-15.4	*24.1*	11.2	173.7	27.4
6	-14.5	*23.8*	12.3	179.8	29.2
8	-13.3	*23.5*	13.5	185.9	31
10	-12.2	*23.1*	14.6	192.2	32.9
12	-11	*22.7*	15.8	198.6	34.9
14	-10	*22.3*	17.1	205.2	36.9
16	8.9	*21.9*	18.4	211.9	39
18	7.7	*21.5*	19.7	218.8	41.1
20	- 6.7	*21.1*	21	225.8	43.3
22	- 5.6	*20.6*	22.4	233	45.5
24	- 4.4	*20.2*	23.9	240.3	47.9
26	- 3.3	*19.7*	25.4	247.8	50.2
28	- 2.2	*19.1*	26.9	255.5	52.7
30	- 1.1	*18.6*	28.5	263.3	55.2
32	0	*18.1*	30.1	271.3	57.8
34	1.1	*17.5*	31.7	279.5	60.5
36	2.2	*16.9*	33.4	287.8	63.3
38	3.3	*16.3*	35.2	296.3	66.1
40	4.4	*15.6*	37	305	69
42	5.6	*14.9*	38.8	313.9	72
44	6.8	*14.2*	40.7	322.9	75
46	7.7	*13.5*	42.7	332.2	78.2
48	8.9	*12.8*	44.7	341.6	81.4
50	10	*12*	46.7	351.2	84.7
52	11	*11.2*	48.8	361.1	88.1
54	12.2	*10.4*	51	371.1	91.5
56	13.3	*9.5*	53.2	381.3	95.2
58	14.4	*8.7*	55.4	391.7	98.6
60	15.5	*7.7*	57.7	402.4	102.5
62	16.7	*6.8*	60	413.3	106.3
64	17.8	*5.8*	62.5	424.2	110.2
66	18.9	*4.8*	65	435.6	114.2
68	20	*3.7*	67.6	447	118.3
70	21.1	*2.6*	70.2	458.8	122.5
72	22.2	*1.5*	72.9	470.7	126.8

Table 3-2 (Cont.)

SATURATED PRESSURE/TEMPERATURE TABLES

Fahr.	R-113	R-114	R-500	R-502	R-717
4	*28*	*16.5*	15.9	35	18.8
6	*27.9*	*15.8*	17	37	20.4
8	*27.7*	*15.1*	18.4	39	22.1
10	*27.6*	*14.3*	19.8	41.1	23.8
12	*27.5*	*13.5*	21.2	43.2	25.6
14	*27.3*	*12.7*	22.7	45.4	27.5
16	*27.1*	*11.9*	24.2	47.7	29.4
18	*27*	*11*	25.7	50	31.4
20	*26.8*	*10.1*	27.3	52.4	33.5
22	*26.6*	*9.1*	29	54.9	35.7
24	*26.4*	*8.1*	30.7	57.4	37.9
26	*26.2*	*7.1*	32.5	60	40
28	*26*	*6.1*	34.3	62.7	42.6
30	*25.8*	*5*	36.1	65.4	45
32	*25.6*	*3.9*	38	68.2	47.6
34	*25.3*	*2.7*	40	71.1	50.2
36	*25.1*	*1.5*	42	74.1	52.9
38	*24.8*	*0.2*	44.1	77.1	55.7
40	*24.5*	0.5	46.2	80.2	58.6
42	*24.2*	1.2	48.4	83.4	61.6
44	*23.9*	1.9	50.7	86.6	64.7
46	*23.6*	2.6	53	90	67.9
48	*23.3*	3.3	55.4	93.4	71.1
50	*22.9*	4	57.8	96.9	74.5
52	*22.6*	4.8	60.3	100.5	76.2
54	*22.2*	5.6	62.9	104.1	78
56	*21.8*	6.4	65.5	107.9	81.5
58	*21.4*	7.3	68.2	111.7	85.2
60	*21*	8.1	71.1	115.6	89
62	*20.6*	9	73.8	119.6	92.9
64	*20.1*	9.9	76.7	123.7	96.9
66	*19.7*	10.9	79.6	127.9	101
68	*19.2*	11.9	82.8	132.3	105.3
70	*18.7*	12.9	85.8	136.6	109.6
72	*18.2*	13.9	89	141.1	118.7

Note: R-114 column values from 16 down to 38 are in. Hg vacuum; R-113 column values from 34 down to 44 are in. Hg vacuum.

Table 3-2 (Cont.)

SATURATED PRESSURE/TEMPERATURE TABLES

Fahr.	Cels.	R-11	R-12	R-13	R-22
74	23.3	*0.4*	75.6	483	131.2
76	24.4	0.4	78.4	495.3	135.7
78	25.6	1	81.3	508.1	140.3
80	26.6	1.6	84.2	521	145
82	27.8	2.2	87.2	534.1	149.8
84	28.9	2.9	90.2	547.5	154.7
86	30	3.6	93.3		159.8
88	31.1	4.3	96.5		164.9
90	32.3	5	99.8		170.1
92	33.3	5.7	103.1		175.4
94	34.4	6.5	106.5		180.9
96	35.6	7.3	110		186.5
98	36.7	8.1	113.5		192.1
100	37.8	8.9	117.2		197.9
102	38.9	9.8	120.9		203.8
104	40	10.6	124.6		209.9
106	41.1	11.5	129.5		216
108	42.3	12.5	132.4		223.3
110	43.4	13.4	136.4		228.7
112	44.4	14.4	140.5		235.2
114	45.5	15.3	144.7		241.9
116	46.6	16.4	148.9		248.7
118	47.8	17.4	153.2		255.6
120	48.9	18.5	157.7		262.6
122	50	19.6	162.2		269.7
124	51.2	20.7	166.7		277
126	52.2	21.9	171.4		284.4
128	53.4	23	176.2		291.8
130	54.5	24.3	181		299.3
132	55	25.5	185.9		307
134	56	26.8	191		315.2
136	57.2	28.1	196.1		323.6
138	58.8	29.4	201.3		332.3
140	60	30.8	206.6		341.3
150	65.5	38.2	234.6		387.2

Table 3-2 (Cont.)

SATURATED PRESSURE/TEMPERATURE TABLES

Fahr.	R-113	R-114	R-500	R-502	R-717
74	*17.6*	15	92.3	145.6	123.4
76	*17.1*	16.1	95.6	150.3	128.3
78	*16.5*	17.2	99	155.1	133.2
80	*15.9*	18.3	102.5	159.9	138.3
82	*15.3*	19.5	106.1	164.9	143.6
84	*14.6*	20.7	109.7	170	149
86	*13.9*	22	113.4	175.1	154.5
88	*13.2*	23.3	117.3	180.4	160.1
90	*12.5*	24.6	121.2	185.8	165.9
92	*11.8*	25.9	125.1	191.2	172
94	*11*	27.3	129.2	196.9	178
96	*10.2*	28.7	133.3	202.6	184.2
98	*9.4*	30.2	137.6	208.4	190.6
100	*8.6*	31.7	141.9	214.4	197.2
102	*7.7*	33.2	146.3	220.4	204
104	*6.8*	34.8	150.9	226.6	210.7
106	*5.9*	36.4	155.4	232.9	217.8
108	*4.9*	38	160.1	239.3	225
110	*4*	39.7	164.9	245.8	232.3
112	*3*	41.4	169.8	252.5	239.8
114	*1.9*	43.2	174.8	259.2	247.5
116	*0.8*	45	180	266	255.4
118	0.1	46.9	185	273.1	263.5
120	0.7	48.7	190.3	280.3	271.7
122	1.3	50.7	195.7	287.6	280.2
124	1.9	52.7	201.2	295	288.7
126	2.5	54.7	206.7	302.5	
128	3.1	56.7	212.4	310.2	
130	3.7	58.8	218.2	318	
132	4.4	61	224	326	
134	5.1	63.2	230.1	334.1	
136	5.8	65.5	236.3	342.3	
138	6.5	67.7	242.5	350.7	
140	7.2	70.1	248.8	359.2	
150	11.2	82.6		404	

Note: For R-113, values from 74 through 116 are in. Hg vacuum.

Table 3-2 (Cont.)

SUPERHEAT

One of the most important fundamental concepts in air conditioning and refrigeration is superheat. Ask ten servicemen to explain it and you may get ten different explanations. Using an electric resistance thermometer, I can usually get an accurate superheat reading on an expansion valve in about half an hour. A young serviceman borrowed the thermometer one day and checked the superheat on eight thermal expansion valves in thirty minutes. I thought he was kidding me but he assured me he had checked each one and the superheat was between 8 and 10 degrees on each valve.

All thermal expansion valves I have ever installed come factory set for 8 degrees to 10 degrees superheat. After installing them and seeing that the remote bulb is properly positioned, making good contact with the suction line and insulated from the air stream if necessary, I usually attach the thermometer and gauges to check superheat. I cannot recall a single occasion when it was necessary to alter the factory setting on the valve. In talking to servicemen, I find that the majority of them consider *adjusting the expansion valve* to be a routine item of business to be done on most installations and service calls. My experience has been that expansion valves do not need to be *adjusted* unless someone has previously done this, then it becomes necessary for me to check for and reset the superheat adjustment.

What is superheat? Look at your pressure/temperature chart, Table 3-2. Refrigerant-12 at 40 degrees F has a corresponding pressure of 37 psig. If you have a suction pressure of 37 psig on a system and the suction line thermometer reads 40 degrees F, you would **not** be correct in saying that the suction gas was coming back at 0 degrees superheat. At zero superheat the suction line can be carrying refrigerant in either a liquid or a vapor state, or a mixture of both. Without superheat you cannot be sure. A temperature/pressure chart gives you the corresponding **saturated** pressure for the temperature. A cylinder of refrigerant is a good example of this. The cylinder contains liquid refrigerant up to a certain level. Above this level the cylinder contains saturated vapor: refrigerant vapor of 100 percent relative humidity, if you wish to put it that way. Start taking away heat and some of the saturated vapor condenses to liquid refrigerant. The pressure is reduced at the same time. Add heat and the vapor expands and drops below its saturation point, some of the liquid vaporizes to bring the vapor back up to 100 percent saturation, the pressure increases. The refrigerant stays in a saturated state.

The expansion valve feeds refrigerant to the evaporator. The evaporator absorbs heat to **vaporize** the refrigerant. We do not want liquid refrigerant leaving the evaporator. We do not want liquid refrigerant entering the compressor. Compressors are made to compress vapor. Liquid is not compressible. To be sure that we have no liquid leaving the evaporator, we **superheat** it. When the refrigerant leaving the evaporator is above the **saturated** temperature corresponding to the pressure, we know it is vapor and not liquid. We know because we have superheated the vapor. Only vapor can be superheated.

We do not want too much superheat because this would mean that we are not getting full utilization of the evaporator. Therefore, we try to keep the superheat at 6 degrees to 10 degrees. How do you find the superheat? You need two things: an accurate suction pressure gauge and an accurate thermometer. On a close coupled job with a short suction line, you can install the gauge on the compressor. Attach the thermometer to the suction line close to the remote bulb of the expansion valve. After the thermometer has settled down, and is fluctuating up and down with the opening and closing of the expansion valve, you can obtain a true reading of the actual temperature of the refrigerant leaving the coil. From this subtract the **saturated temperature** corresponding to the suction pressure. The difference is the superheat.

The catch is in getting an accurate temperature reading on the suction line and in analyzing what it means. With a good resistance thermometer on the suction line, you may find the temperature fluctuating as much as 20 degrees as the valve opens and closes. It is best to watch the thermometer for at least 15 minutes before making any change in valve adjustment. Usually, a continuing wide fluctuation means the superheat setting is too high. Before making any change decide what you are going to do. Superheat setting on a thermal expansion valve is a balance between the pressure of the gas in the remote bulb and diaphragm trying to push the valve open and the spring in the valve body trying to keep it closed. If you want a lower superheat, you want less spring pressure. Turn the adjusting stem back ¼ turn, then observe the temperature fluctuations for at least 15 minutes. It takes time for the valve to settle down and operate at any new setting. Make it a rule never to change the adjustment more than ¼ turn at a time, and wait 15 minutes between changes. It may seem to take longer but you will get an accurate setting much quicker this way.

An accurate suction pressure reading is as important as an accurate temperature reading. With long suction lines, heat exchangers, many fittings or several expansion valves on one condensing unit,

you will find that a reading taken at the compressor is almost useless. In this case you must get a gauge reading fairly close to the
remote bulb location. If the expansion valve has an external equaliz-
ing line, you can put a line tap valve on this line. Or you can pump
down the unit and silver braze a Schrader valve in the suction line.
Quite a few contractors are teeing a valve in the equalizer lines as
part of all new installations these days. It pays off in good service
and customer satisfaction at very little initial expense.

For the most part, only three things can cause thermal expansion
valve trouble: sticking due to dirt, sludge or moisture; not opening
due to a lost charge in the power element; and improper superheat
setting.

Some servicemen claim that it is more accurate to use two thermometers and read the difference between the refrigerant entering
the coil and the refrigerant leaving the coil as superheat. This will
hold true **only if the pressure is the same at these two points.** Since
there is internal friction loss in every coil, I do not see how this method of reading superheat could be accurate.

Use good instruments and take accurate readings. Take time for
the valve to settle down to work after each change. Above all, **think.**
You will be surprised at how much better performance you get from
a system after the superheat has been properly set. Once you have
made the proper setting **never** change it unless you have proved the
need for the change first by taking an accurate reading of the superheat.

SUBCOOLING

Subcooling is just the opposite and yet similar to superheat in air
conditioning and refrigeration. By superheating the vapor leaving
the evaporator just a few degrees above the saturated temperature
corresponding to the pressure, we insure that we are getting full use
of the evaporator and that no liquid is returning to the compressor.
By subcooling liquid leaving the receiver just a few degrees below the
saturated temperature corresponding to the pressure, we insure that
no flash gas is forming in the liquid line, and we are getting full use
of the expansion valve.

True subcooling and superheating have one thing in common, a
change in temperature but not in pressure. The amount of refrigerant that can feed through the expansion valve depends on the difference in pressure across the valve. Flash gas forming in the liquid line
will reduce the refrigerant flow. Subcooling will prevent flash gas.
However, too much cooling will reduce the pressure in the liquid line

and reduce flow through the valve.

Normally, we do not worry about loss of liquid line pressure when we are subcooling by means of a liquid-suction heat exchanger. The pressure of the hot gas in the condenser is transmitted through the liquid to maintain full pressure in the line. Subcooling on evaporative condensers and air-cooled condensers is usually accomplished by passing liquid leaving the receiver through a few turns of tubing located in the incoming air stream before leaving the condenser.

I have heard servicemen complain about too much subcooling causing loss of head pressure and thus starving the evaporator and causing shutdown on low suction pressure. With manual reset control systems and lockout relays, this has resulted in a good many nuisance calls. To be correct, this is not due to subcooling. This is loss of condenser pressure and usually occurs on systems that are operated in low outside ambient temperatures. The problem usually occurs after an extended shutdown that has allowed the refrigerant in the condenser to become chilled to a point where it has insufficient pressure to push the required amount of refrigerant through the expansion valve to maintain the suction pressure. You may have noticed that the problem is much more common on Refrigerant-22 than on Refrigerant-12. Normally, Refrigerant-12 head pressure is 125 psi, suction is 40 psi and pressure drop across valve is 85 psi. Refrigerant-22 head pressure is 212 psi, suction is 69 psi and pressure drop across valve is 143 psi. You can see why loss of head pressure is more critical on R-22.

There are various methods used to maintain head pressure on today's condensers. For air-cooled systems, the most common is the pressure valve which traps liquid refrigerant in the condenser and reduces the condensing area. This is fairly effective but it does have its disadvantages. A large amount of refrigerant is required. A large receiver is required. The evaporator is temporarily starved during liquid buildup. Operation is fairly effective providing the system can be kept in constant operation. In the case of large tonnages, opening of the valve can sometimes produce hydraulic surge which can damage the liquid line unless it is firmly anchored against such surging. Some air-cooled condensers use head-pressure-operated inlet air dampers to maintain head pressures by controlling the quantity of condenser air. Others, with multi-fan condensers, use modulating or step controls to cycle the fans.

I have seen a drawing of a proposed air-cooled condenser which had four sections and four fans. It is to be used on a four-step capacity controlled compressor and as each step cuts out, one section of the condenser was to be closed off by means of solenoid valves on the inlet and outlet lines, and its fan cut off. In theory this will work. I question its economic feasibility. The more complicated the control system becomes, the more troublesome and expensive the operation becomes.

With the scarcity of good refrigeration servicemen today, we should keep our installations as simple as possible. A refrigeration contractor in Kansas recently showed me his solution to keeping an air-cooled condenser operating even in below freezing temperatures. He built an insulated room around the condenser. An inlet damper low in one wall and an exhaust damper high in another wall were controlled by a modulating thermostat sensing the inlet air to the condenser. Thermostat was set at 75 degrees and he claimed fairly good operation the first winter. The only problem was starting up after a long off cycle when the room had a chance to cool off too much. He added an electric heater controlled by a thermostat set at 60 degrees and claimed very good operation through the next two winters. The room had two feet clearance on each side of the condenser and four feet at front and back, and removable doors at each end for summer operation. It seems like a lot of work and expense to keep the system operating, but the owner was very happy with it, and said it had paid for itself in reduced service costs.

One installation I serviced a few years ago was a modern drug warehouse. This was an automated setup: overhead conveyor racks, selective switching of trucks to the proper loading dock and storage area, packages shunted off of the belt conveyors by scanner eyes, computerized inventory control and billing. The air conditioning was a 200-ton hermetic centrifugal with a cooling tower and two gas-oil-fired scotch boilers for a two-pipe hot water system. Four air units handled the warehouse and one took care of the office area. The computer room was in the center of the office area and had its own 7½-ton air-cooled system. This system was primarily for use before the main system was started in the spring, and after the fall shutdown. Needless to say, we had quite a bit of trouble with it. It had an electric refrigerant heater to maintain head pressure on the off cycle, but with off cycles that sometimes lasted for a week, it could not do the job properly.

Since this unit was primarily a peaking unit with limited use, I have often wondered why the condensing unit was not installed in the warehouse instead of on the roof. The heat rejected in the winter would have helped the warehouse and the added load in the summer would have had very little effect on the large system. The 75 degree to 80 degree average ambient air available to the condenser would have eliminated the need for any head pressure controllers, and cut the maintenance costs considerably. I cannot think of a simpler solution to this problem.

DESUPERHEATING

Desuperheating is not new. It has been around for some time under the name of liquid injection. Injecting liquid refrigerant into the heads is common practice on some large compressors to remove heat that is not moved out during periods when the cylinder is unloaded. It is necessary to keep the injector assembly in top operating condition because it can be very damaging when it goes wrong.

Manufacturers are getting away from liquid injection by designing compressors so that some gas always moves through unloaded cylinders to the remaining loaded cylinders. You should never find a compressor that will unload every cylinder. At least one should always be loaded.

It is one thing to inject liquid on the discharge side and quite another to inject it on the suction side. We keep TXV superheat at least 6 degrees to insure that we do not slug the compressor. We even install accumulators as antislug devices. The serviceman is still going to run across installations where suction gas needs to be desuperheated in order to keep hermetic compressors operating.

The simplest method I have found for this is shown in Fig. 3-1. The solenoid valve is wired to open when the compressor comes on. Use a capillary tube for a ¼-hp freezer unit on compressors from 1 to 3-hp. Use a capillary tube for a 1/3-hp freezer unit on 3, 4, and 5-ton compressors.

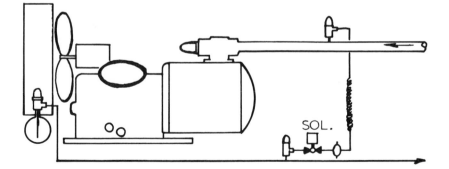

SOL.

FIG. 3-1

On larger hermetic compressors, use a TXV feeding a ten foot length of ½-in. tubing that is coiled and insulated Fig. 3-2. Set the superheat at 10 degrees, never less. Most hermetic compressors on air conditioning service will get enough cooling if the suction gas enters at between 55 degrees and 60 degrees. On commercial installations, take the temperature of the discharge line where it leaves the casing or read the head temperature directly. Temperatures between 180 degrees and 200 degrees are what you want for best operation.

Above all, **think.** Most hermetic compressors take the suction gas over the motor first and can vaporize a little liquid here. The small capillary tubes or TXV's will not slop over too much, if at all. Installing a capillary tube designed for a window unit, or too large an expansion valve, could allow enough slopover to tear a compressor to pieces. Installing shutoff valves on the desuperheater assembly will allow you to make changes if your thermometer indicates you need them.

Before you start installing a desuperheater, check the superheat. Make sure you have a full charge of refrigerant. Make sure the suction line is insulated all the way. If the system has ever had a burnout, or if the superheat check indicates erratic operation, check the expansion valve for dirt or sludge. Installing desuperheating is not going to be the answer if any of these items are causing trouble.

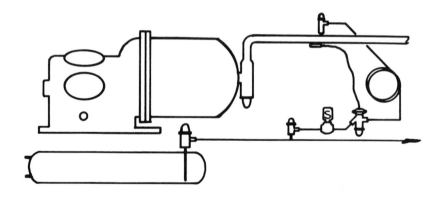

FIG. 3-2

4

Head Pressure

We have been discussing the laws of gases and how they apply to refrigerants in order to understand what happens inside a refrigeration system. So far we have been concerned only with the operation of a system that contains a single gas such as Refrigerants-12, 22, or 717 (ammonia). What would happen if a refrigeration system contained two gases?

There is a law of gases that covers the situation. It is known as Dalton's Law and is stated as follows: the total pressure of a mixture of gases is the sum of the partial pressure of each of the gases in the mixture. When you have a mixture of gases in a system, pressures, temperatures, and resultant actions are not going to be very predictable because the end results are going to be very different from those of each individual gas. There are exceptions to this and we will get into them a little later on.

If you have a mixture of air and refrigerant in a system you are going to have high head pressures. This is due in part because of Dalton's Law, but mostly because air is noncondensable in a refrigeration system and upsets the normal operating balance. A refrigeration compressor will pick up and pump air as part of the natural flow in almost all systems. The air becomes trapped in the condenser because it requires pressures far beyond the compressor's capabilities to condense it into a liquid. Thus, the air remains in a gaseous state and takes up space that is normally used to remove heat from the refrigerant. The net result is to reduce the condensing capacity of the system and, in turn, cause higher condensing pressures. Higher condensing pressures mean that the hot gas enters the condenser at higher temperatures. This means a greater TD between the gas and the condensing medium. A greater TD accelerates heat transmission and the system will eventually reach a state of balance where the heat rejected by the condenser equals the heat absorbed by the evaporator. Total capacity will be reduced, operating costs will increase, but the system will operate. It will operate somewhere between its design condition and the limits set by the high limit safety switches.

Eventually, as air volume increases, the machine will cut out on these safety switches.

When the serviceman is confronted with the problem of high head pressures he has the problem of finding out just what is causing it. On the present hermetic systems air in the system is seldom the cause of high head pressures. It is not as infrequent on process refrigeration systems operating at low back pressures, provided they are open-type compressors. Automotive air conditioners are often air-bound because they usually lose gas over the winter and are then operated for a while without gas before they come into the shop. I will never recharge one of these systems without first evacuating it.

Insufficient air over the condensing coils of an air-cooled unit could be another reason for high head pressure. It could be caused by worn belts, worn motor sheaves, or oil and dirt-encrusted blades on either a propeller-type fan or a squirrel-cage fan. Bent fins and dirt imbedded deep between the fins cut down on air flow and also insulate against heat transfer. Fin combs are available to straighten fins and you can brush off surface dirt, visible to your eye, that may be blocking air flow.

Dirt imbedded deep between the fins is a problem. Tubes are

FIG. 4-1   Condenser cleaning tools.

usually staggered so it is difficult to see straight through these coils. It is possible to put a flashlight on one side of the coil and catch a glimpse of the light through the other side. This will be some help in finding out if there are areas of the coil that are dirty enough to block air flow. Another thing you can do is apply detergent solution to a small area of the coil and wash it off after five minutes with a water hose. If you can see for yourself that the cleaned area is visibly cleaner, you can be sure that the fins have enough dirt on them to insulate against heat transfer. I don't know how accurate it is, but I have heard that .002 in. of dirt on the fins will cut down heat transfer by 10 percent. My own experience tells me that air-cooled condensing units in residential areas in the midwest need to be cleaned every two years. Units in downtown and industrial areas need cleaning every year.

There are a variety of tools available to the serviceman for detergent cleaning of condensers. The hand pump spray bottles can be used on small units and are very handy for testing small areas on large condensers. You will often find that the dramatic difference between cleaned and uncleaned areas is the only sales tool you need to sell a condenser cleaning job. For the larger systems I use a portable diaphragm-type air compressor together with an Imperial Brass Co. jet spray gun. The gun has a venturi for drawing detergent solution into the air stream and sprays very deep into the coil. Its disadvantage is that you must be careful in your choice of detergents. Some of them, when fogged into the air, will drive you out of the work area. The jet-type washers that attach to a garden hose and aspirate detergent into the water stream are something new, and have turned out to be very good. Cost is low and performance is good. You may have to repeat the detergent spray once or twice in order to clean all the way through the coils, but the results are worth the effort.

Whatever method you use, be sure you do a good job of cleaning all the way through the coil and of washing off all the detergent solution. Failure to do a thorough cleaning job could leave chemicals that would attack the mechanical bond between the primary surface of the copper tube and the secondary surface of the aluminum fins. Corrosion of this vital bond can result in a drastic loss in condensing capacity.

Atmospheric pollution, in the form of sulphur dioxide and carbon dioxide fumes washed from the air by rain, can lodge in the dirt on coils and cause crevice corrosion. This has been a problem mainly in units installed in industrial areas. It is now becoming more and more common with widespread pollution. Keeping coils as clean as possible is the best answer to this growing problem.

If the serviceman has convinced himself that the problem of high head pressure is caused by air in the system, the next step is determining how to remove it. The design of the system, type of refrigerant, condensing medium, must all be taken into consideration when air must be removed by the serviceman. An air-cooled condenser usually has a small header for hot gas entrance, with very little space for collecting air. If purge valves are installed they are usually at the top of the coil on the end opposite the hot gas entrance. With the system off, air will percolate up through the tubes to the top and can be bled off through the valve. Purging usually must be done several times in order to do a good job. A careful check of operating pressures between purging sessions will tell whether you are getting the air out.

If the system is small and the refrigerant charge not too large, compare the cost of purging versus a complete blowing of the charge, then pulling a vacuum and recharging. Evacuating/recharging is the most positive method of air elimination and takes less time. It may be the least costly method.

Water-cooled condensers of the tube-in-tube type present about the same problem in purging as the air-cooled types, except for one added hazard. If purging continues to the point where liquid refrigerant boils in the condenser at temperatures below freezing, you can freeze the condensing water and burst a tube. This is not an unusual thing to happen, so **think** when you are purging water-cooled condensers.

Shell-and-tube condensers are the easiest to purge because they have a large space available to trap the air. Purge valves are usually installed for this purpose. Combination shell-and-tube condenser receivers are also the most susceptible to freeze-ups and burst tubes during purging because they contain liquid refrigerant in contact with the lower water tubes. Watch the pressure gauge carefully during purging!

Centrifugal systems usually have purge units installed as original equipment. These purge units have a compressor that draws air and refrigerant vapor from the top of the condenser and pumps it to a small condenser in the purge unit. Liquid refrigerant from the purge condenser is fed back to the system by a float valve and air is vented to the outside by a pressure relief valve.

Since air taken from the top of the condenser also contains any water vapor that may be present in the system, the purge unit is sure to be the place with the most corrosion. The purge unit requires the most maintenance. It must be kept clean and be checked frequently to make sure it is operating correctly. It is a good idea to install a drier on the liquid line leaving the purge condenser to clean this refrigerant.

Commercial refrigeration systems, especially those using ammonia (717), often employ purge units that use a refrigerated coil to condense the refrigerant. The refrigerated coil creates a low pressure area in the unit, eliminating the need for a compressor to draw air and refrigerant vapor to the unit. Pressures are low enough so that the unit can draw from evaporators as well as condensers. This is important because ammonia can break down under some conditions and create foul gases that must be removed. Purging is done manually, as needed, and the gases are vented into a bucket of water to absorb traces of ammonia gas. Cleaning maintenance to remove corrosion is still required in these units.

Remember the old belt-driven compressors with the cast fins on the upper part of the cylinders, the cast iron heads with the fins that collected all that oily dirt, the water-cooled valve plates?

Why the fins and water-cooled plates? To get rid of the heat of compression buildup in the compressor! When you compress refrigerant vapor to raise the temperature so that the heat may be rejected to the condensing medium, you have to go to a higher temperature than the condensing medium. This heat is also transmitted to the metal of the compressor. Superheated vapor entering the compressor at 60 degrees is superheated to 180 degrees or 200 degrees as it leaves the discharge valve. Raise the entering superheat and you also raise the leaving temperature. It's the law (of gases).

If you have done much commercial refrigeration work you can probably remember cases where the vapor entering the crankcase was cold enough to cause frost to form yet you could burn your hand on the discharge head. Ammonia compressors are still made with water-cooled jackets on the heads. The greater amount of heat carried by ammonia makes this necessary.

Heat must be removed from motors and compressors or it will cause damage. Valve discs, strips, springs will lose their spring temper and fail to seat. Heads will warp and blow gaskets between the high and low sides. Motor insulation will fail because of heat and bearings will distort and lubrication fail. Compressor oil must be kept cool because this is what removes heat from the piston. **Think** about it. Where else can piston heat go?

The refrigerant vapor returning to the compressor must always be cool enough to remove heat from the machine. Some of the small domestic refrigerator compressors have a short finned tube to aid in oil cooling. Two-stage systems normally have intercoolers to assure that vapor discharge from the first stage is cool enough to remove heat from the next stage compressor.

The fact that you have a frosted crankcase does not mean that you automatically have enough cooling for the compressor. It simply means that the vapor is returning at a temperature below freezing. Low temperature gas is low pressure gas and low pressure gas is low density gas. Low density gas cannot pick up nearly as much heat. A compressor operating on a low back pressure is going to be the first one to give discharge valve trouble because of overheating.

When hermetic compressors came into being, heat removal became a real problem. Suction gas had to be piped to the motor chamber first to carry off the heat. This meant that the gas entering the compressor was superheated even more than was usual. Valve plates started giving trouble. Some manufacturers put out information bulletins that limited the size of heat exchangers for liquid line subcooling. They also recommended **against** the common practice of taping or soldering the liquid and suction lines together for subcooling purposes. Insulating the suction line all the way to the compressor was advised to prevent additional superheating after leaving the evaporator.

One compressor manufacturer installed a water coil in the compressor to keep the oil cool. Fine, as long as you are working with above-freezing suction temperatures. It's a real source of trouble if you forgot to blow all the water out of this coil before the winter shutdown on an air conditioning installation. Many a compressor motor has grounded for this reason.

The fact that hermetic compressors depend on suction gas for cooling brings a new problem for the serviceman. He must make sure that the TXV superheat is not too **high.** He must also see that further superheating of the suction gas is held to a minimum. And when he finds an installation where he still does not have sufficient compressor cooling, he has to come up with an answer.

5

Psychrometrics

When a serviceman understands the laws of gases and applies this knowledge to what is happening inside the system, it would seem that he has all that it takes to make a proper diagnosis and determine the need for repairs. This is not always true. In air conditioning and commercial refrigeration a knowledge of psychrometrics is necessary to understand what is happening to the atmosphere that is being acted upon by the refrigeration or air conditioning system.

As an example of how a knowledge of psychrometrics would have helped a serviceman: The call was on a residential air conditioning system that just would not cool the house to the thermostat setting. It had done the job satisfactorily in the past in the same weather conditions, so something must be wrong with it. The serviceman put gauges on the compressor and read the temperature of the air leaving the cooling coil.

Air was leaving at 64 degrees so he decided to lower this temperature. The evaporator coil had an adjustable thermostatic expansion valve on it, so he took off the cap and turned the valve stem in one turn. He saw the back pressure go down a few pounds and reasoned that this would give him a lower coil temperature and a colder leaving air temperature. He noticed the evaporator fan motor had an adjustable sheave, so he stopped the fan and moved the sheave in one-half turn to speed up the fan. The combination of a colder coil and a little more air over the fan should do the job or so he thought.

The customer called back the next day and said it was worse instead of better, so the serviceman was sent back out. He called in and asked for a replacement compressor. The back pressure was up instead of down and doing a lot of fluctuating, so he reasoned that he had compressor trouble. Since the system was in warranty, the service manager sent me out to double-check.

When I walked in the temperature was too high for real comfort, but what I noticed most was the sticky feeling of the air that said high humidity to me. The serviceman told me what he had done and why he had decided the compressor was bad. He was a good kid and really wanted to learn, so we went through the service work step-by-step to see what we could do.

The wide fluctuation of the suction gauge indicated high super-heat, and the fact that he had turned the TXV stem in meant he had increased the superheat. The first thing we did was to turn the valve stem back one turn and attach a thermocouple lead of my resistance thermometer to the suction line leaving the coil. The next step was to take an ammeter reading of the fan motor and compare it to the motor nameplate rating. The motor was fully loaded. Next, we looked at the furnace air filters. They were clean in color but had a heavy layer of lint on the surface. We replaced them and rechecked the motor amperage. It had moved up just a little, so we set the adjustable sheave back down to take care of this.

I explained to the serviceman that turning the TXV stem in had lowered the back pressure. It had also raised the superheat and decreased the capacity of the coil to absorb heat. We checked the suction line temperature and compared it to the saturated temperature corresponding to the suction pressure and found we were operating at about an 18 degree superheat. We agreed this was too high and backed off the valve stem one-half turn. While we were waiting for the TXV to settle down, we walked around the basement and I noticed the washer and drier in the laundry were operating. I asked the serviceman if they had been operating the day before, and he said they had been.

We rechecked the superheat and found it was now varying between 8 and 10 degrees. This was an excellent superheat reading, so I got out my sling psychrometer, Fig. 5-1, moistened the wet bulb wick, and took readings at two of the supply registers. Air was leav-

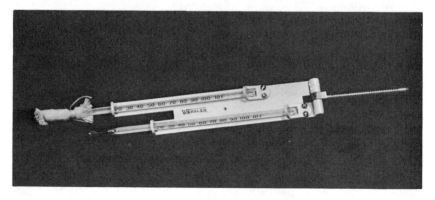

FIG. 5-1 Sling psychrometer.

ing the registers at 65 degrees dry bulb and 62 degrees wet bulb. The high dry bulb could be accounted for by heat picked up traveling through the ducts in the basement, but the high wet bulb indicated we were trying to take out more moisture than the coil was designed to handle.

I asked the lady of the house if she was doing more laundry than usual and learned that her month-old twin daughters were keeping the laundry very busy. To make a long story short, we sold her on letting us take out a pane of glass in a basement window near the drier and piping the vent from the drier out through a metal plate installed in place of the glass. We placed a screened cover over the window well to catch the lint. The service manager called in a few days later and asked how things were going and found they were happy and comfortable. The serviceman asked me what books to read to learn about psychrometrics.

Psychrometrics can be defined as: the study of the physical and thermodynamic properties of the atmosphere.

In air conditioning, heating, and refrigeration the properties we are concerned with are: dry bulb temperature, wet bulb temperature, dew point temperature, specific humidity, relative humidity, sensible heat, latent heat, total heat, density, and pressure.

Air is a mixture of gases and the atmosphere is composed of these gases plus water vapor and dirt, dust, and other pollutants put into the atmosphere. Air itself is a mixture of 78 percent nitrogen, 21 percent oxygen, and 1 percent argon, neon, carbon dioxide, sulfur dioxide and other gases. The amount of water vapor varies constantly depending on changes in temperature and location of the air. The warmer the air the more moisture it can hold; the cooler the air, the less it can hold.

Dry bulb temperature is the temperature of the air taken with a thermometer with a dry bulb. Wet bulb temperature is taken with a thermometer with the bulb covered with a cloth wick saturated with water, preferably distilled water. Air passing over the wick will cause part of the water to evaporate as the water follows the laws of gases and tries to balance its vapor pressure with the vapor pressure of the water vapor in the air. Moist air will have a high vapor pressure and will not evaporate much from the wick. Dry air will have a low vapor pressure and will evaporate a lot more water from the wick because of the greater difference in pressure. We are operating under the laws of gases and as this moisture evaporates it will use up heat energy in the water and lower the temperature of the water in the wick.

It follows, then, that the drier the air the lower the wet bulb temperature will be. Wet and dry bulb temperatures are obtained with a sling psychrometer, Fig. 5-1, an instrument that has a wet and a dry bulb thermometer mounted on it. The sling psychrometer is whirled through the air to obtain a reading that is representative of the total air mass and not just the air immediately surrounding it.

When you have obtained the wet and dry bulb temperatures, you can determine humidity. Humidity is used in two ways in our business. One is specific humidity which is the actual weight of the moisture in the air in terms of grains of moisture per pound of dry air, or pounds of moisture per pound of dry air. There are 7,000 grains per pound, so the use of grains weight takes a lot of decimals out of the figures. The other is relative humidity which is the percentage of moisture vapor present in the air as related to the amount it could hold at the actual temperature of the air.

Specific humidity does not change as temperature of the air changes. If a pound of dry air contains 100 grains of moisture at 80 degrees, it will still hold 100 grains at 100 degrees. Relative humidity does change as the temperature changes. Remember that the warmer the air the more moisture it can hold. 80 degree air that contains 100 grains of moisture is 65 percent saturated. Heat this air to 100 degrees and it will still contain 100 grains of moisture, but the relative humidity will have changed to 35 percent.

Specific humidity is important in designing air conditioning systems and in checking the amount of work being done by the equipment. Relative humidity is an important measure of the conditions produced by the equipment.

The human body is a machine. It does work and, in order to do this work, it uses energy. This energy comes from food that the body burns. Part of the energy obtained from food is in the form of heat. The body rejects this heat by radiation and the evaporation of perspiration. The drier the air around the body, the faster this perspiration is evaporated.

One function of air conditioning is to maintain the relative humidity at a level where evaporation of moisture from the body is fast enough to keep us comfortable. The serviceman must be able to measure the relative humidity in conditioned areas. To do this he will use a sling psychrometer to obtain the wet and dry bulb readings. He then subtracts the wet bulb temperature from the dry bulb temperature. The result is called the wet bulb depression. Referring to the relative humidity tables, Table 5-1, locate the wet bulb depression at the top and the dry bulb temperature on the left side. Where these lines cross is the relative humidity.

The psychrometric chart can also be used to find relative humidity. Locate the dry bulb temperature at the bottom. Dry bulb temperature lines run straight up. Locate the wet bulb temperature on the curving line to the left. Wet bulb lines slant down to the right. These two lines will intersect somewhere on or between the percent relative humidity lines that curve up to the right. You can estimate the percentage between the lines with fair accuracy.

The intersection point you have just found will be on or close to a horizontal line. Follow this line to the left and you will reach the 100

PERCENT RELATIVE HUMIDITY TABLE

Wet Bulb Depression

Dry Bulb Degrees F	1	2	3	4	5	6	7	8	9	10	12	14	16	18	20
30	89	78	67	56	46	36	26	16	6						
32	89	79	69	59	49	39	30	20	11	2					
34	90	81	71	62	52	43	34	25	16	8					
36	91	82	73	64	55	46	38	29	21	13					
38	91	83	75	66	58	50	42	33	25	17	2				
40	92	83	75	68	60	52	45	37	29	22	7				
42	92	85	77	69	62	55	47	40	33	26	12	2			
44	93	85	78	71	63	56	49	43	36	30	16	7			
46	93	86	79	72	65	58	52	45	39	32	20	11			
48	93	86	79	73	66	60	54	47	41	35	23	15			
50	93	87	80	74	67	61	55	49	43	38	27	20	5		
55	94	88	82	76	70	65	59	54	49	43	34	26	14	5	
60	94	89	83	78	73	68	63	58	53	48	40	32	21	13	5
65	95	90	85	80	75	70	66	61	56	52	44	37	27	20	12
70	95	90	86	81	77	72	68	64	59	55	48	42	33	25	19
72	95	91	86	82	77	73	69	65	61	57	49	43	34	28	21
74	95	91	86	82	78	74	69	65	61	58	50	45	36	29	23
76	96	91	87	82	78	74	70	66	63	59	51	46	38	31	25
78	96	91	87	83	79	75	71	67	63	60	53	47	39	33	27
80	96	91	87	83	79	75	72	68	64	61	54	49	41	35	29
85	96	92	88	84	81	76	73	70	66	62	56	49	44	38	33
90	96	92	89	85	81	78	74	71	68	65	58	52	47	41	36
95	96	93	89	85	82	79	75	72	69	66	61	55	50	43	38
100	96	93	89	86	83	80	77	73	70	68	62	56	51	46	41

Table 5-1

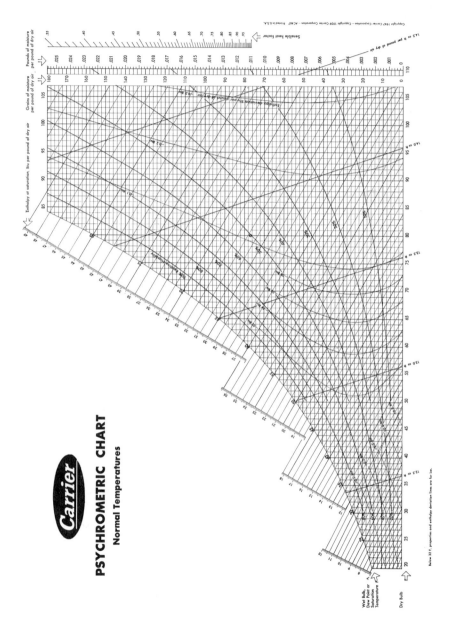

PSYCHROMETRIC CHART
Normal Temperatures

FIG. 5-2　(Reprinted by permission of Carrier Corporation.)

percent relative humidity line. The temperature there is the dew point temperature. When you cool the air below this temperature, you will take some of the moisture out of the air. How much will you take out? Follow the same line to the right and you will have the specific humidity of the air. Take a sling psychrometer reading of the air after you have cooled it below the dew point. Locate the intersection of the new wet and dry bulb readings, and again follow the horizontal line to the right. You will then have the new specific humidity reading. The difference in weight between the two readings will enable you to figure the actual weight of the moisture that has been removed from each pound of dry air.

The lines slanting from the bottom up to the left are air density lines. The warmer the air, the lighter it is, and the greater volume of air it takes to make up a pound. If the temperature of the air is 85 degrees and the relative humidity is 50 percent, the density of the dry air is 14 cu. ft. per pound. If our air conditioning system moves 14,000 cfm of this air, it means we are moving 1000 pounds of air per minute plus the weight of the moisture in the air.

A temperature of 85 degrees and 50 percent rh (relative humidity) gives an absolute humidity of 90 grains per pound of dry air. 1000 pounds X 90 equals 90,000 grains. At 7000 grains per pound, we are moving 12.9 pounds of water vapor per minute with the 1000 lbs. of air. Do you still wonder why fan motors, belts, and pulleys wear out?

Sensible heat refers to the heat you can actually measure with a thermometer. Latent heat is the heat released when you condense the water vapor to a liquid. You cannot measure it with a thermometer but if you will take the weight of the water removed, the temperatures at which you removed it, and refer back to the tables on the properties of water vapor and find the latent heat at this temperature and pressure, you can total up the amount of heat that had to be removed from the air to do this job.

The engineer designing the system to do a certain amount of work will be using the enthalpy tables at the left side of the curving line. Enthalpy is the combination of temperature and pressure changes that enable the engineer to find the energy required for the job. He will find the two intersections of the air before and after being conditioned, and project a line through these points up to the right. There the sensible heat factor table helps to select the proper type of evaporator coil for the installation.

A serviceman with the knowledge of psychrometrics can certainly do a better job of servicing than the man did on the residential job we last read about. There is no reason why he cannot take care of a service call on a large commercial air conditioning system.

Psychrometrics is equally important in refrigeration work. Any evaporator coil that cools a refrigerated room moves air in some fashion. It does not have to be a blower coil. An overhead bare-pipe coil sets up convection currents in the room air that effectively cool all the air in the room, provided the coils are not blocked with ice, frost, or boxes and cartons stacked on them. **Think about it.**

As an example of the importance of psychrometrics in commercial refrigeration: Some years ago a man walked into my office and asked me if I would be interested in selling him a refrigeration system. When I told him I most certainly would, he gave me the details. He was operating a small slaughterhouse in western Kansas and supplying meat to restaurants in the surrounding area. His chill room was too small for the load and he had built a new one. He gave me the dimensions, type of construction and average weight and number of head he killed. I figured the load and gave him a price. He left, saying he was getting prices from other people too. When he came back two days later, he had some questions to ask. Why was I $800 higher than other contractors and why did I insist on using two blower coils where the others were only using one?

I explained to this man that coils were rated by the Btu's absorbed per degree TD (temperature difference) and that I had tried to select the coils for his job on a 10 degree TD between refrigerant and room temperature. I had picked a 10 degree TD because I wanted to keep the relative humidity in the room high to reduce meat shrinkage. Also, a small TD meant a higher operating back pressure and this, in turn, meant that the compressor was more efficient because it was pumping a denser refrigerant gas and actually moving more heat per revolution. I must have gotten my reasons across to him, because he bought the job. I installed it and made a trip out there every 60 days, for the year after, to honor my warranty. He never did call for any service between my calls and, on my last warranty trip, he showed me some sides of beef he had been aging and asked me to guess how long they had been in the room. The fuzz on the beef was about ⅜ in. thick and since I was holding a 31 degree room and maintaining a defrost cycle, I guessed at 30 days. I was wrong. They had been there 62 days. He told me that my story on the proper TD for selecting coils and its relation to meat shrinkage had so impressed him that he had kept accurate weight records on his beef in and out of the rooms. According to his calculations, the savings on shrinkage in the new room was enough to pay the entire cost of the refrigeration installation in the first year.

That is one man who knows how important relative humidity is in a commercial refrigeration system.

A walk-in refrigerator for storing flowers requires a high relative humidity and a fairly high holding temperature, about 50 degrees on the average. Warehouse storage of photographic films and papers are at about 50 degrees. So are lead-zinc storage batteries and pharmaceuticals. You must maintain a lower relative humidity for these products than you do for flowers. If you do not you will have container damage from moisture and, in some cases, fungus and mold growth that will cause product damage.

The serviceman confronted with a problem of high humidity in a commercial cooler is going to have to do some thinking to come up with a solution to the problem. Before you do anything to the machinery, check out the condition of the cooler walls, door gaskets, etc. Regardless of whether you are dealing with a domestic refrigerator or a large cold storage warehouse, exterior moisture is always trying to get into the lower temperature area. It is just obeying the laws of gases.

If the problem is one of equipment that is not specifically designed for the products stored in the cooler, figure your solution in a logical manner. Less air over the coil will mean both a lower coil temperature and more dehumidification of the air. If reducing air volume will cause a problem in room air coverage, bypassing part of the air around the coil will reduce humidity. Partly covering the discharge face of the coil with metal strips, spaced so that they cover about one-third of the area, will make the air stay longer in the coil and be dehumidified more. The air will be forced between the strips at a higher velocity and have almost the original volume. You can find the answer to your problem if you think about it.

REHEAT SYSTEMS

When a customer calls in and says that the air conditioning system won't run, you can send your serviceman out with explicit instructions, "Find out why it won't run and fix it." This may or may not be an easy job, but at least the serviceman has a definite starting point to work from. He knows the system is supposed to run, he can find out why it is not running, and he can make the necessary repairs to get it running again.

What if the customer says, "Everything seems to be running but we get complaints. Different areas are too cold or too warm or the air is too dry or too humid, and there does not seem to be any fixed pattern to these complaints. Can you send a man over?" Sure you can send a man over, but can you give him explicit instructions? All you can say is, "It's not working right. See if you can find out what is wrong and fix it."

Now you are on the job and you are looking for a place to start. What do you do? See if everything is running! Good! We will see if everything is running. And while we are doing this, we will also be **thinking,** what is it? What does it do? What is its part in the system design?

Walking into the return air plenum chamber we start looking around. What do we see? Opposed blade dampers in front of the return air duct. Another set of opposed blade dampers the same size in the side wall. What does this mean to you, two return air dampers? This is possible, but what about one return air and one outside air damper?

Let's see if the air temperatures match room air and outside air temperatures. They do and it is logical to have a fresh air intake on a system of this size. Why is it the same size as the return air? Are there times when you would use all outside air? We are trying to maintain 75 degrees in the conditioned space. If the outside air is below 75 degrees, it can be used instead of return air and the system will operate at a lower load. Now, are the dampers operating correctly? They are opposed blade dampers with modulating operators. This means they are supposed to mix outside air and room air. The outside air damper is open about 20 percent and the return air damper about 80 percent. This would be right if there were a minimum fresh air position switch in the control system. We can check this and we can also check the damper linkage to make sure it is tight and not slipping when the operator moves. Everything seems to be okay, but wait! When the system changes to outside air automatically, where does the return air go? There should be an exhaust damper in the return air ductwork system to allow this air to leave the building. One more damper to find and check for correct operation before we move on. The next step is the filters. We check these to make sure they are in the frames, not plugged with dirt and not bypassing air. This done, we move on to the fan. Belt tension okay, sheaves not worn or loose on the shaft, no bearing noise, fan blades clean and free of dirt and grease.

Next we find finned steam heating coils. The steam valve is closed and since it is summer, this seems normal. We check the coils to make sure they are clean and not restricting the air flow, and with everything looking okay, move on. The DX coils are next. The coil face has a little dust and lint on it, and this is wiped off with a rag.

The tubes are not bright so we know there is some dirt between the fins and on the tubes. The coil return bends are all cold and the condensate is running off of the fins and down the drain in a fairly heavy stream. Stepping out of the air handling unit access door, we find four thermal expansion valves and four liquid line solenoids. It is evident that we have evaporator capacity control.

Moving out to the compressors, we find two water-cooled condensing units, hermetic compressors with internal unloading capacity control, and shell and tube condenser/receivers. The gauges on the control boxes indicate about 220 lbs. head pressure, 60 lbs. suction pressure and 150 lbs. oil pressure. Since these are mounted gauges and subjected to the operating vibration of the compressors, we have reason not to trust these readings. For the time being, we will accept them as near correct. The suction and liquid lines *feel* to the hand is about right.

Returning to the air handling system, we find several air ducts leaving the system and manual balancing dampers in these ducts to regulate the amount of air each zone duct handles. A steam heating coil is in each duct after the balancing damper. Checking through the access doors, we find these coils fairly clean and the steam valves for each one closed. So far so good. We haven't checked the cooling tower or the pump, but since the head pressure gauge reads okay, we have no indication of anything not operating correctly. Why should there be complaints?

Before we start adjusting or changing any thermostats or controls, let's stop and think about this job. What kind of a system is this? It has steam coils to preheat the air entering the unit. It has DX coils to cool and dehumidify the air. It also has steam coils to reheat the air. This is a **reheat** system. If you are going to find out what is wrong with this system, first you will have to know what it should do and how it should operate. We must know the purpose of a reheat system.

A reheat system is designed to dehumidify the air to a certain specific humidity, and then supply whatever sensible heat is needed to maintain the space temperature at design conditions. Let's take a look at this on a psychrometric chart, Fig. 5-3. Point A is design room conditions. This is the point of intersection of desired room wet bulb, dry bulb and relative humidity. A horizontal line drawn through this point will intersect with the saturation curve on the left and the specific humidity scale on the right. In order to deliver air at X grains of moisture to the room it is necessary for us to cool it to point B. We now have air at the right specific humidity. In order to have the right

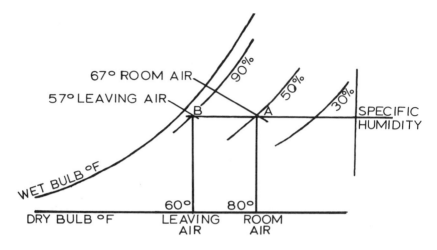

FIG. 5-3   Reheat psychrometrics (taken from Psychrometric Chart).

relative humidity, we must add sensible heat along line B-A. Part of this sensible heat will be added when the air leaving the grilles mixes with and picks up heat from the room air. If the room temperature goes below the design temperature, the thermostat will call for re-heat to bring the room temperature up to point A.

The purpose of a reheat system then, is to cool and dehumidify the air to a certain specific humidity. The air is then reheated as needed to maintain the room temperature and relative humidity.

Armed with this knowledge, the serviceman can now do some logical reasoning. Cold and damp complaint? Check the reheat or maybe someone has turned off the steam because it is summer. Not enough dehumidification? What are the wet bulb and dry bulb temperatures of the air leaving the DX coil? If they are too high, then we can look for the specific cause. It may be shortage of refrigerant, dirty coils giving poor heat transfer or even the compressor loading control set for the wrong operating back pressure. Is the air too dry? Once again you can look for the specific cause. Put your own gauges on the com-pressor. If the back pressure is too low, the evaporator temperature will be too low and the leaving specific humidity will be too low. You will have a starting point to look for the cause of this trouble because you know the purpose of the system design, and you can **think.**

6

Refrigerants and Controls

Before you start applying your knowledge of the laws of gases, know what refrigerant is in the system. A quick glance at an old reference book shows about 75 different refrigerants listed in the numbering system. This list started with No. 10, carbon tetrachloride. Normally, it is used as a cleaning fluid and fire extinguishing fluid. About the only time you will find it used as a refrigerant is in a laboratory or other special purpose use. The list ended with No. 1270, propylene, a by-product of catalytic cracking of petroleum products. It is a good refrigerant, having a boiling point of minus 54 degrees F at atmospheric pressure, but you will usually find it used in the refineries as a basis for the production of such items as isopropyl alcohol, acetone, detergents, hydraulic fluids and lubricants.

This is not an unusual thing to find with refrigerants. Teflon, the nonstick coating for cooking utensils and also used for gaskets, pipe thread tapes, valve diaphragms, is made by adding a monomer to the familiar Refrigerant-113 used in centrifugal refrigeration machines.

A lot of refrigerants have almost entirely disappeared from general use. Sulfur dioxide (R-764) was at one time the most widely used refrigerant for domestic and light commercial refrigeration. When electric refrigeration came into being, we used to install a finned coil with compartments for the ice cube trays in the ice compartment of the icebox and run the copper tubing to a compressor in the basement. Charged with sulfur dioxide, it would do a beautiful cooling job compared to using ice.

Sulfur dioxide is highly absorbent of moisture and was commonly used as a drying agent for dried fruits for this reason. Mixed with water, it became sulfurous acid, H_2SO_3. In the system, it carbonized the oil and deposited this carbon on the interior surfaces. It was toxic and when it leaked it could and did kill insects . . . also pets, plants and people. It has been a long time since I serviced SO_2 systems and I hope I never have to service one again.

Ammonia, NH_3, is Refrigerant-717. It boils at -28 degrees F and was the first widely used commercial refrigerant. It is still in wide use, although it does not get enough publicity to seem so. Most ammonia service is done by plant maintenance personnel because there are not too many qualified servicemen left in this field. Those who are in the field say they do not lack for work.

Ammonia is toxic. It irritates the eyes and nose and can cause serious burns. It is explosive when mixed with air in two definite mixtures and under certain conditions. It is flammable any time it is spewing from a ruptured pipe and carrying oil vapor with it. This is true of any refrigerant carrying vaporized oil along with it. Do not open electrical switches or permit any sparks or open flames near a leak that is spewing oily vapor. Cut off the power from an exterior source and fog the area with a water or steam spray to reduce the fire hazard.

I have worked on ammonia systems for the last 35 years and still carry my gas mask and sulfur tapers in the truck. In some respects, it is an easy refrigerant to work with. You can always tell when you have a leak and it is relatively easy to hold ammonia. Since the only refrigerant that carries more heat per pound of refrigerant is water, I think ammonia will be used for a long time to come.

LP gases are also refrigerants and one, isobutane, was used in domestic refrigerators for a while. Its fire hazard was too great for it to become popular. Methyl chloride, R-40, was a common refrigerant for a while around World War II. With a boiling point of -10 degrees F it was not too far off from R-12 and was used as a substitute when R-12 could not be obtained. It is flammable if enough is present in the air. R-40 dissolves aluminum and the resultant product is tri-methyl-aluminate. TMA is highly flammable. Exposure to air is all that is necessary to touch it off. The flame is so intense that it is comparable to a thermite welding process.

R-40 was responsible for a good many fires that I know of and one very funny incident. A multistage centrifugal converted from R-12 to R-40 in a war production plant gradually lost capacity. It finally reached the point where it was turning just as many rpm's as ever, but it was not producing any cooling at all. When it was opened up all they could find was a bare shaft and the two steel setscrews in the bottom of the casing. Even the FBI could not find out *who stole the impellers?*

Fluorocarbons are the refrigerants the serviceman will find in most systems today. They are made by a number of companies and put out under different trademarks. Freon is the trademark of du Pont; Isotron of Pennwalt; Ucon of Union Carbide; Genetron of Allied Chemicals. R-12 is R-12 regardless of who makes it and as long as you use the same R number you can put the refrigerant in the machine.

Fluorocarbons are not toxic and for this reason are safe to use in our refrigeration and air conditioning systems today. Adequate ventilation is necessary around the equipment as fluorocarbons will not sustain life if no oxygen is present. Fluorocarbons that have been through a fire or in contact with hot copper elements such as a halide torch or while soldering or brazing tubing, break down and form highly toxic gases. Phosgene is the most common of these. Always provide adequate ventilation when leak testing or using a torch.

R-11 and R-113 are commonly used in centrifugal systems. R-11 boils at 74.8 degrees and R-113 at 117.6 degrees at atmospheric pressure. R-114 boils at 38.8 degrees and is especially suitable for domestic refrigerators using rotary compressors such as Frigidaire's *meter-miser*. R-13 has a boiling point of -114.6 degrees and is used as the secondary refrigerant in cascade systems for extremely low temperatures. The condenser of the R-13 system is cooled by the evaporator of the primary machine. If this evaporator is operating at a 40 degree temperature, the R-13 condensing pressure is only 305 psig.

The four most commonly used refrigerants in air conditioning and refrigeration are: 12, 22, 500, and 502. 500 is an azeotrope, 48.8 percent 22 and 51.2 percent 115 by weight. 502 is an azeotrope, 73.8 percent 12 and 26.2 percent 152a by weight.

The serviceman must determine the refrigerant in the system before he can add refrigerant or use a pressure temperature chart to set superheat or know if the gauge pressures he is reading are correct. Most nameplates give the type and amount of refrigerant in the system. There is a possibility that the refrigerant has been changed from the nameplate designation in some cases. To double check, look at the expansion valves. They should name the refrigerant and be color coded. Color codes are: yellow for R-12, green for R-22, orange for R-500, purple for R-502, and white for R-717. R-40 is red and codes for 11, 13, 114, 764 are blue. These are the same colors as the factory-filled refrigerant cylinders.

Some condensing unit manufacturers use color coded tags on the machines. In this case watch out. Early Carrier units used red for R-500. R-500 was developed by Carrier and used and manufactured exclusively for them until the patent ran out.

Each refrigeration system uses a specific refrigerant for a definite reason. It may be the temperature range of the product being cooled or it may be the design of the refrigeration machinery. The fact that so many refrigerants are available enables the manufacturer to use

fewer sizes of compressors to produce many units with different capacities. The combination of compressor, motor and condenser, as
assembled by the manufacturer, is designed to pump so much gas per
revolution. This gas carries a designed amount of heat. If the serviceman changes the refrigerant to another number, he is changing the
unit capacity. A lot of engineering work must be done to prove that
such a change is possible and to see what changes to the equipment,
controls, etc. must accompany such a change. Expansion valves must
be changed, or at least have new power elements and orifices installed. Safety and operating controls must either be changed or reset to different pressures. Line sizes must be checked to see if they
are sized for the different volumes of liquids and gases to be carried.

REFRIGERANT CONTROLS

We have already talked about superheat and how to find it and set
an expansion valve. Thermostatic expansion valves, Fig. 6-1, operate
by a pressure difference between the spring trying to hold the valve
closed and the power element trying to force the valve open. Pressure
in the power element depends on the temperature of the gas or liq-

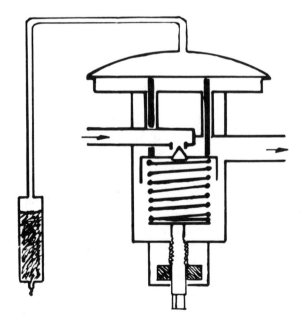

FIG. 6-1 Thermostatic expansion valve.

uid-gas charge in the power element, tube, bulb. Ideally, the temperature of the bulb is the effective force. To insure that this is so, the bulb must make good contact with the suction line and be insulated from the effect of the surrounding air. Check this before you check the superheat. You may find this is the answer to erratic coil operation.

It is possible for a thermal expansion valve power element to completely or partially lose its charge. With a complete loss of charge, the valve will close and changing the spring pressure has little or no effect. A partial loss of charge is not as easy to detect. The usual symptom is a wide, sluggish, fluctuation in suction line temperature. Before you decide this is the answer, make sure the reason is not dirt or sludge in the valve. My experience has been that a valve operates erratically from dirt and sludge in it at least ten times more often than if it has a partial charge loss.

Don't be too quick to blame some troubles on the expansion valve. I was once asked by a department store manager in St. Louis to check on a contractor's recommendation that 16 TXV's be replaced. He was having trouble keeping the back pressure down and getting enough dehumidification and temperature drop across the coils. He reasoned that the valves were old and worn and not closing tight.

The contractor's serviceman and I checked the job thoroughly and found that the two 12-cylinder compressors were having trouble maintaining oil pressure because of worn bearings. Low oil pressure was making it impossible for the oil-pressure-operated unloader pistons to keep the back pressure down. After rebuilding the compressors, the system operated perfectly. The expansion valves operated okay and were still doing the job at an 8-10 degree superheat setting almost 10 years later.

Another time I received a call from a store manager in Nebraska for help. His serviceman had replaced the expansion valves on a 30-ton air unit twice and still could not get the back pressure down. I was familiar with the job and so, just on a hunch, I asked the manager to take a look at the compressor end bell to see if the one-eighth inch threaded opening below the acorn nut of the capacity control was still open. The capacity control is actually an automatic expansion valve adapted for this purpose and the opening is one that maintains atmospheric pressure on top of the valve diaphragm. Sometimes the gasket leaks where the valve is bolted to the inside of the end bell and you will detect refrigerant at this opening. If a serviceman does not know the purpose of this opening, he would be tempted to plug it. Refrigerant pressure would build up on top of the diaphragm and affect valve operation.

This turned out to be what had happened. After the plug was removed, the machine immediately loaded up and brought the pressure down. The leak from the opening was stopped by tightening the pressure on the gasket with the two allen-head cap screws that hold the control in place.

The only remaining problem was what to do about the bill from the serviceman for the expansion valves. As a supervisor charged with seeing that the store got the service it needed and did not spend money needlessly, I could not see paying for new valves and work that need not have been done if the serviceman knew his equipment and was **thinking.** I made my recommendation to the store manager to this effect, but the final decision was up to him. It always is in a case like this. He is the man responsible for operating the store and showing a profit in that store and in that town. He wrote me later that he had paid for two valves and still had a good serviceman to take care of his equipment. And, the serviceman was more knowledgeable about the equipment than he had been. Now that is for sure!

Thermal expansion valves (TXV) maintain superheat by converting temperature of the refrigerant leaving the evaporator coil to pressure operating the diaphragm that opens and closes the valve needle. Temperature depends on the pressure in the suction line. Pressure on the bottom of the diaphragm will oppose the bulb pressure. Some TXV's have internal equalizers to bleed pressure from below the diaphragm to the low side. The larger a coil is, the more restriction to refrigerant flow it will have. To operate properly, the pressure on the bottom of the diaphragm should be the same as the pressure in the suction line. The large capacity TXV's have external equalizers. These are fittings that allow a small line, usually one-quarter of an inch, to be run from the valve to a fitting in the suction line. You may find the equalizer line is brazed into a drilled hole in the suction line. I have run into a few cases where this tube was not far enough through the hole when it was brazed and the braze plugged the end of the line.

The serviceman with a valve that will not settle on a reliable superheat setting might check this possibility. Kinked or flattened equalizer lines will have the same effect. The equalizer line is a perfect place to get a pressure reading for superheat work. Attach a line tap valve to the line close to the suction line.

There are many other refrigerant controls besides TXV's. Automatic expansion valves (AXV), Fig 6-2, were widely used in the days of brine or alcohol ballasted coils. You will find them on low sides that operate at a more or less constant load. An AXV is simply a spring-controlled valve that maintains a constant pressure on the leaving side. They are ideal for maintaining a constant suction pressure providing load is constant. You do not worry about setting superheat on these controls as much as you do about keeping the back pressure at a certain setting.

The capillary tube is the most widely used refrigerant control on small refrigeration units and window air conditioners. The length and diameter of the tube, in comparison to the pressure differential across it, determines how much refrigerant passes through it. Setting the superheat depends on getting the proper charge of refrigerant in the system. You can obtain the superheat reading just as in any other control. You change the superheat by adding or removing refrigerant from the unit. On domestic food freezers, you add refrigerant to give a definite frost length on the suction line. This can have its drawbacks. If the unit is loaded with food and at normal operating temperatures inside and in the normal temperature room, you can be fairly certain of the proper charge. Change these conditions and you may over or undercharge.

The simplest way, in most cases, is to evacuate the system and add the preweighed charge or use a charging unit that allows you to measure the proper charge in a visible tube. Some manufacturers supply detailed instructions as to what pressures to charge to at different ambient air conditions. These instructions are precise and result in proper charging **only if the serviceman reads them care-**

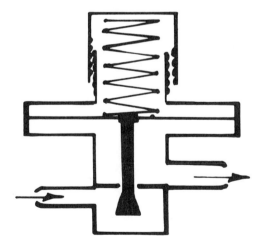

FIG. 6-2 Automatic expansion valve.

fully and then follows them to the letter. This sounds like a small item but it is actually the biggest cause of trouble in service work.

High side floats, Fig. 6-3, were used in domestic refrigerators before the capillary tube and are still used in commercial refrigeration and air conditioning systems. The serviceman will see them on the centrifugal machines he services. As the name implies, the float is in the high side of the system. The valve controlled by the float is the point of separation between high and low sides.

Refrigerant that collects in the float chamber raises the float and feeds into the low side. Superheat on high side float systems is controlled by the amount of refrigerant in the system. When I was working as a factory branch serviceman I had a regular monthly circuit of installations to inspect and maintain. Most of the larger air conditioning systems were sprayed coil low sides that used high side float refrigerant control. The compressors were belt driven or synchronous direct drive motors with shaft packing instead of the shaft seals of today. A little refrigerant leakage at the packing was normal. A superheat check and addition of R-12 to bring the superheat down between 6 degrees and 10 degrees was part of each month's work.

Maintenance problems on high side floats usually are due to needle and seat wear from foreign matter and wire drawing. A system that was not clean and dry could also cause corrosion and damage to the float. Early model Trane centrifugal systems have a sight glass in the float chamber that allows observation of the float and its action. It is certainly a great help in servicing these units.

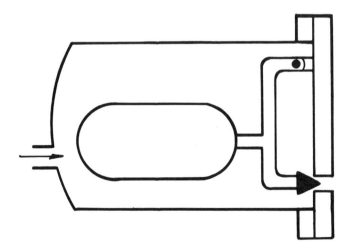

FIG. 6-3 High side float.

Low side floats, Fig. 6-4, as the name implies, have the float in the low side of the system. They maintain a set level of refrigerant in the evaporator. Superheat is determined by the level of the float. Float assemblies are piped into the level designated by the factory and most of them have an adjustment stem or screw that allows a variation to be made in the level at which the float closes the needle valve. On some makes this is an internal adjustment and requires that the float assembly be valved off and evacuated before opening the float chamber to make the adjustment. The serviceman would want to be very sure of the need for adjustment before making this change. The Phillips low side float has sight glasses and an external adjusting stem. I have found a lot of these on Vilter Mfg. Co. installations. The serviceman will find that setting superheat is a lot easier with this type of float valve.

A fairly new development in refrigerant control is typified by the Level Master Control of Sporlan Valve Co. Basically, it is a thermal expansion valve. The remote bulb is not strapped to a suction line but is inserted in a special well that has an electric heating element in the bulb jacket. Liquid refrigerant around the bulb carries the heat away faster than it can heat the bulb. This closes the valve. When the liquid level drops, the heat reaches the bulb and opens the valve. Limited superheat adjustment is possible from the valve adjustment stem. I have found some of these controls installed with a remote bulb and well chamber piped in with flexible hoses. Changing the height of

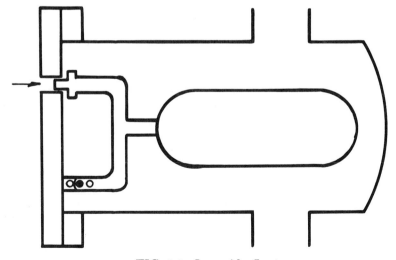

FIG. 6-4 Low side float.

this chamber changes the liquid level in the evaporator. Level Master Controls are especially susceptible to oil around the bulb affecting operation.

Refrigerant controls are precision devices and must be kept clean and free of corrosion in order to function properly. Liquid line strainers and driers are a part of almost every refrigeration system. Moisture indicators are common on almost all new installations nowadays. These accessories are so valuable to the proper operation of the system that they should be a part of every installation.

Not long ago, I was sent out to service an installation that consisted of four 60-ton air-cooled DX water chillers. Moisture indicators had been installed on each system and a drier in a liquid line bypass. No strainers. Two of the compressors had suffered internal breakage because of a design fault that allowed the chillers to load up with refrigerant while the compressors were off. Slugging damage created a lot of debris and with no strainers a lot of it lodged in the cages of the Alco TXV's. When I installed new cages I also installed line strainers ahead of the valves. Saving money on an installation by not installing strainers ahead of the refrigerant controls is an expensive saving later on.

All fluorocarbon refrigerants should be kept clean and dry. It is possible to install strainers, driers, and moisture indicators on the average system because these items are designed with this in mind. What is the serviceman going to do when he finds that he has a problem with moisture, corrosion or oil contamination, on something like a centrifugal system that does not have the liquid and suction lines we are used to. **Think** about it!

Most centrifugals have a purge unit which is used to remove air and moisture from the refrigerant. The purge unit is a compressor and condenser that pulls vapor from the top of the condenser and compresses it. The compressed gas is condensed and a float valve returns the liquid to the system. Water collects on top of the liquid and is removed by manually opening a blowdown valve as needed. Air and other noncondensable gases are removed by a spring-loaded pressure relief valve.

The water that collects on the surface of the refrigerant is excess moisture above that which the refrigerant itself will absorb. While the absorbed moisture itself probably will not cause freeze-up in the system, it will contribute to the formation of acid in the refrigerant. The reason I put driers on a centrifugal is primarily to remove acid. Acid can damage the float valves. It can cause crevice corrosion at tube and tube support sheets. It can reduce the dielectric strength of motor winding insulation on hermetic centrifugals. I have known a hermetic centrifugal burnout to cost over $9,000.00 to clean out and replace motor. Preventive maintenance on centrifugals is a must.

It is not possible to install a full-flow liquid line drier on a centrifugal nor is it necessary, since the prime purpose of the drier is to remove acid. I have found that all the centrifugals I have worked on have thermometer wells or plugs or valved openings which make it possible to install a drier in a bypass line around the float valve. Since the capacity of a centrifugal can be varied to as low as 10 percent of its rating, the bypass around the float must be only 10 percent of the flow at maximum.

The idea of a bypass drier is to clean up a small percentage of the refrigerant and return it to the system where it will reduce the total contamination. Eventually, the whole refrigerant charge will have the contamination greatly reduced.

I have used a replaceable core drier with a 2 in. opening. This opening has been bushed down to ½ in. tube for systems under 200 tons. For 600-ton systems, I have used a ¾ in. tube. I always install a sight glass with a moisture indicator on both sides of the drier. This gives you a chance to see the liquid flow and the moisture indicator works on R-11 and 113 just as they do on 12 and 22. The percentages will not be the same but the manufacturer will supply you with this information if you ask for it.

On some centrifugals I have had to punch out the bottom of a thermometer well in order to install a valve in the sump at the bottom of the condenser. Some makes have threaded thermometer wells that can be removed and extended with a tee with a valve in the side. Some makes have a pipe plug in the sump. Whatever the make of machine, I always try to pipe liquid refrigerant from ahead of the float valve, through the drier, and return it to the evaporator at a low point.

Since a drier will do its best when in contact with liquid refrigerant, I always put a valve or a riser, or both, in the line leaving the drier to maintain a liquid column in the drier. If I use a riser, I also put in a bypass to drain the drier of liquid when the cores are to be changed.

Some of the centrifugal purge units have a ¼ in. copper line to return the liquid from the purge condenser float valve to the system. A small drier installed in this line, with moisture indicator on the leaving side, will insure that the returned refrigerant is dry. I have also installed driers in the suction lines of the purge unit. The idea here was to remove some of the acid from the vapor entering the purge unit and reduce corrosion and wear in the purge compressor itself. I find that it does help, although it certainly does not eliminate the need for periodic cleaning and oil changing of the purge compressor.

I have also installed activated alumina core driers in the oil lines. Oil manufacturers use activated alumina to reduce acidity in oil. It serves the same purpose here. I have to be careful on this installation as oil will not flow through the driers unless it is heated to 125 degrees. An oil pressure interlock switch is standard on most centrifugals and I make sure that this switch is connected after the drier in case it should stop the oil flow.

Another form of contamination of refrigerant that is found on centrifugals is oil that gradually collects in the evaporator and reduces the efficiency of the system. When it gets too bad you can either send the refrigerant back to the manufacturer to be reprocessed or you can dump the charge and put in a new charge. The refrigerant manufacturers will analyze a sample of refrigerant and tell you just how much contamination is present.

I have set up a distillation unit on several installations that has helped prolong the useful life of the refrigerant. I welded a coupling on the side of a 200 lb. refrigerant drum near the bottom. This opening was connected to a valve at the low point of the evaporator. A line was run from the top of the drum to a connection on the suction line of the centrifugal. Using a boiler sight glass and a two inch union less one tailpiece, I made a sight glass for the two inch drum opening.

Before starting up the system, the valves are opened and refrigerant is allowed to flow into the drum until it is about half full. The inlet valve is then closed and when the system is operating the suction causes the refrigerant to distill off. At the end of the day's operation the drum is valved off and the residue of oil and sludge can be poured from the drum. It is a slow process but, where a plant has operators available, it is possible to keep the system remarkably clean by use of this simple daily procedure.

It is almost impossible for me to overstress the need for **thinking** when the serviceman is working at getting the proper superheat on a system using float valve refrigerant controls or with flooded systems. The amount of damage that liquid slugging can do to a machine is incredible. You have to open up a machine and find all the bits and pieces of valve plates, pistons, cylinder liners, connecting rods, and wrist pins piled up in the crankcase to appreciate it.

The old slow-speed compressors had spring-loaded valve plates that would lift when slugged. The noise a plate made when it lifted was quite unnerving. Today's high-speed machines do not have the spring-loaded plates. They would be of no value at the high speed. Flooded brine chillers now have vapor domes to minimize liquid carry-over, Fig. 6-5. Centrifugals are built with eliminator plates in the evaporators for this purpose, Fig. 6-6. Suction accumulators are installed for antislugging protection.

On a recent ammonia installation, we had brine chillers as well as low temperature flooded blower coils with a separate rotary booster. The brine chiller had a vapor dome. The 8-cylinder direct drive 600 rpm compressor slugged one day and scared the wits out of everybody in the place. The trouble turned out to be a low side float ball that had broken off from the stem. We did not run the compressor until we had checked it out for internal damage.

Slugging a centrifugal compressor can be extremely damaging. A 30 in. diameter impeller turning 3550 rpm has a tip speed of about

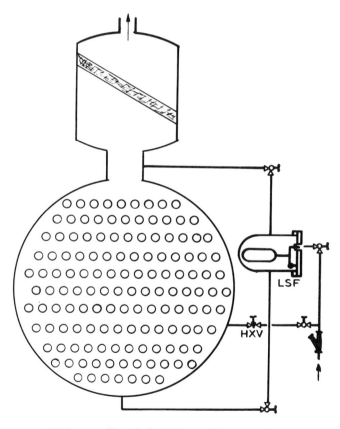

FIG. 6-5 Flooded chiller with vapor dome.

300 miles per hour. At this speed drops of liquid refrigerant would have about the same effect as buckshot hitting the aluminum impellers.

Most chillers have sight glasses or some means of determining the liquid level in the chiller. Find out what the liquid level is supposed to be and do not exceed it. Superheat is not as important as safe operation of the equipment. Think twice, or even three times, before you make any changes that would lower superheat on these machines.

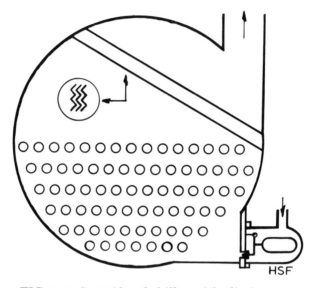

FIG. 6-6 Centrifugal chiller with eliminators.

7

Water and Corrosion

The most abundant single substance in our world is the very familiar, highly unusual, inorganic compound called water. Try as he will, the serviceman cannot escape contact with it. If you only work on air-cooled systems you have to contend with the effects of rain water on the condenser and condensed vapor on the coil and drain pan. If you only work on gas furnaces without humidifiers, you still have to contend with the effect of water vapor on the heat exchanger and highly acid water condensate in the flue under certain conditions. If you work up a sweat just thinking about it you still have to contend with that form of moisture.

Water has many peculiar attributes. It can exist in nature in three states: vapor, liquid, and solid. It can exist simultaneously in these states, as you can see when you look at a pond with ice floating on it and fog swirling above it. Like everything else, including ice, water shrinks when it is cooled, but water stops shrinking at four degrees Celsius and expands from there to the freezing point. This is the reason that ice floats on top of water and is also the reason for burst water pipes and loss of refrigerant in water chillers with improperly set low-limit switches.

Water has a great many thermal properties. It has the highest specific heat known among liquids, meaning it can store more heat energy for a given increase in temperature than any other liquid. Water has the highest latent heat of vaporization among liquids. At 20 degrees C (68 F) it takes 585 calories to evaporate one gram of water (1054 Btu per pound). And, with the exception of mercury, water has the greatest thermal conductivity of all liquids.

These are the reasons why water is used in heating systems and chilled water systems . . . why water is such an excellent condensing medium and is so adaptable for use in cooling towers and evaporative condensers. Pound for pound, no other liquid can do so much energy-carrying work in air conditioning, heating and refrigeration.

There are disadvantages. The high freezing point is one of them. You can lower the freezing point by adding substances to water, but you lower the thermal properties when you do this. Water will dissolve almost anything to some extent. Fortunately, for our purposes, this extent is extremely small in many substances.

Water has the greatest known dielectric constant of any substance. The dielectric constant governs the ability to hold dissolved substances in suspension. Because of this, liquid water is very seldom chemically pure. Water vapor is always pure and ice can easily be made pure. For these reasons liquid water is an ionic solution and among the ions are hydrogen ions. They are always present because the water itself can so easily supply them.

The degree of dilution of liquid water by hydrogen ions gives us a means of measuring the state or condition of water. Percentage of hydrogen ion dilution is expressed as pH. The chemist with a pH meter can rate pH from 0 to 14. The serviceman with litmus paper can rate pH fairly accurately from 2 to 11. A pH of 7 is approximately neutral. Readings below a pH of 7 indicate water of acidic nature and readings above a pH of 7 indicate alkaline water.

The amount of dissolved solids held in suspension in the water is rated as the *hardness* of the water. Hardness ratings may be given as parts per million (ppm) or as grains per gallon. One grain per gallon is equal to 17.1 ppm. Metals held in suspension are principally calcium, magnesium, and sodium. They are not in pure form but as compounds. The type of compound may be as a bicarbonate, carbonate, sulphate, or chloride. To understand the why of these suspended solids, let's take a quick look at the cycle of water.

Liquid water anywhere in the world is evaporated into the atmosphere because of the changing temperatures and pressures. The evaporation process follows the laws of gases. Water vapor is pure water. When this water vapor is precipitated from the atmosphere to the ground, it washes the air of other solids and gases that are present. Rain water reaching the ground is slightly acidic because it has picked up carbon dioxide (CO_2) and sulphur dioxide (SO_2) gases: H_2O plus $SO_2 = H_2SO_3$, H_2O plus $CO_2 = H_2CO_3$.

This acidic water travels over and through the earth until it reaches the streams and wells we use. Because of its acidic nature, it dissolves and carries with it particles of the substances it contacts in its journey. Water treatment plants clean the water by using a lime-soda process. The lime-soda process causes metals such as iron and copper to precipitate out in settling basins. The water is then filtered, treated as necessary to reduce pH to approximately neutral, and chlorinated to kill bacteria.

The amount and type of water treatment varies with the location in the country and source of water. There are many places where well water, as pumped from the ground, requires no treatment at all. Water from heavily polluted rivers requires a great deal of treatment. The water problems that the serviceman will run into will vary with each locale.

For example: Well water used for cooling a department store in Austin, Minnesota, did not affect the heat transfer ability of the cooling coil as tested after 12 years without internal cleaning of the coil. A similar installation at Bloomington, Minnesota, required annual internal coil cleaning. A 40-ton R-22 condenser on a cooling tower at New Ulm, Minnesota, showed a loss of capacity due to tube fouling at the end of the season. The serviceman cleans the tubes by removing both heads and allowing the tubes to dry for at least 30 days. As the deposit in the tubes drys, it turns to a fluffy white coating. Brushing with a nylon bristle brush removes this and leaves a clean, bright copper surface.

At Jefferson, Iowa, and Nevada, Missouri, water conditions cause encrusted buildup in the cooling towers that cuts tower capacity way down during the cooling season. Much as I hate to do it, I had to have water softeners installed on the supply lines to both of these installations. Installations at Cape Girardeau and Springfield, Missouri, require lengthy acid cleaning procedures for condensers that are way out of proportion to the chemical analysis of the water supply. Both of these towns have large cement and lime plants south of town. My experience tells me that is the reason for the problem. In Lenexa, Kansas, a Carrier centrifugal installation with cooling tower had only a minor acidizing cleanup of the condenser each year until the sixth year. That was the year they installed a batch plant, for the new interstate highway construction, a mile and a half southwest of the building. Cooling towers and evaporative condensers are exceptionally good air washers.

Before, I could add between 20 and 30 pounds of inhibited dry acid and circulate it six hours. A pH test at this time would have shown that the solution was still acid at a pH of about 4. Tube inspection showed them to be clean. This time I used 200 pounds and it took two full days of circulating to get the tubes clean.

There are two kinds of water hardness, temporary and permanent hardness. Water with temporary hardness contains bicarbonates. Bicarbonates are metals combined with an acid radical. Calcium bicarbonate is $Ca(HCO_3)_2$. Heat a bicarbonate and it breaks down. Hydrogen ions and carbon dioxide are released. Because this is an electrochemical action, the calcium carbonate that is left from this action plates out on the nearest surface following electrochemical laws.

Water with temporary hardness is the kind that causes the scale formation so common in hot water heating systems and refrigeration condensers.

Water with permanent hardness contains carbonates. Calcium carbonate is $CaCO_3$. Because the electrochemical action involved in its formation is already completed, it does not plate out so easily in boilers and condensers.

The serviceman must understand that all of these compounds present in water are made up of elements in combinations. These elements have the various combinations of protons, neutrons and electrons peculiar to them. They have molecular energy and are subject to the same laws that govern gases. Anytime we use water to carry energy for us in an air conditioning, heating, or refrigeration system, we are going to get some kind of reaction because of this fact. Knowing this will help us in logical thinking about the causes and cures for water problems in our work. One of these problems is corrosion. Let's see just what corrosion is.

CORROSION

Each element has its individual atomic structure. Most of the atoms that make up an element are identical. The principle difference in the atomic structure between elements is in the number of electrons and the pattern of electron movement around the proton. An electron can be considered as a unit of energy. The more electrons there are in an atom, the more energy that atom has. The more movement the electrons have, the more energy the atom has.

Elements, therefore, differ in the amount of energy they contain and, because of this difference, try to exchange electrons until energy is in balance. To accomplish this, the elements must be in a media or solution that is agreeable to transporting the electrons. Water, preferably water that has a low pH number, is the ideal electron transfer solution. There are many chemical compounds that will help in electron transfer. Water, because of its high dielectric constant, is the ideal solvent in which to suspend these compounds.

This is the electrochemical process that we call corrosion. This is what has happened when the bottom of the cooling tower rusts out. We may swear at it when we look at the rust holes in the rocker panels of our car, but we are thankful for it when we turn on a flashlight and get light to see our way on a dark night. Corrosion is a process that can be very helpful to us when we know how to use it.

Since energy variation between the elements is in electrical quantities, it is logical to assume that it is possible to chart the potential force available between the elements, Table 7-1.

The information in these tables is basic material for the electrochemist. Using this tool, he can change or modify energy forces. The development of such new things in our life as the alkaline battery, and space vehicle energy cells, comes from basic information such as this.

The serviceman studying and working in this service field is going to run across words whose definitions he may not be sure of. Words

STANDARD OXIDATION POTENTIALS
AT 25 DEGREES C

ELEMENT	SYMBOL	REACTION PRODUCTS		E°, VOLTS
		Ion	*Electrons*	
Lithium	Li	Li+	1-	3.05
Potassium	K	K+	1-	2.93
Calcium	Ca	Ca++	2-	2.87
Sodium	Na	Na+	1-	2.71
Magnesium	Mg	Mg++	2-	2.37
Uranium	U	U+++	3-	1.80
Aluminum	Al	Al+++	3-	1.66
Manganese	Mn	Mn++	2-	1.18
Zinc	Zn	Zn++	2-	.76
Chromium	Cr	Cr+++	3-	.74
Iron	Fe	Fe++	2-	.44
Cadmium	Cd	Cd++	2-	.40
Nickle	Ni	Ni++	2-	.25
Molybdenum	Mo	Mo+++	3-	.20
Tin	Sn	Sn++	2-	.14
Lead	Pb	Pb++	2-	.13
Hydrogen	H	2H+	2-	0.00
Copper	Cu	Cu++	2-	-.34
Silver	Ag	Ag+	1-	-.80
Mercury	Hg	Hg++	2-	-.85
Palladium	Pd	Pd++	2-	-.99
Platinum	Pt	Pt++	2-	-1.20
Gold	Au	Au+++	3-	-1.50

Table 7-1

such as: anode or anodic, cathode or cathodic, noble, base, passivity. Let's get them straight now so we can make the best use of what follows.

An electrode is a conductor through which current can enter or leave a conducting medium. It can be the steel wire that produces the spark to ignite an oil burner, or it can be the welding rod held in the arc welder's stinger, or it can be a minute speck of impure metal in a steel plate where corrosion starts. An electrode can be anodic or cathodic, depending on the direction of current flow. A cathode is a point of entry for electrons from an electrolyte; an anode is a point of departure for electrons from an electrolyte. Remember, we are dealing only with direct current in these electrochemical actions. Polarity is reversible, anodic can become cathodic depending on polarity, but the current is always direct current.

Noble metals are usually those below hydrogen in the electromotive force series. Base metals are those above hydrogen. Base metals will usually lose electrons to noble metals. The terms noble and base are also used to refer to the relative position of two or more metals in a compound or electrically linked device. Nickle is a base metal compared to hydrogen. Nickle is a noble metal compared to magnesium. Copper is a noble metal coupled with nickle, but copper is a base metal coupled with silver or gold.

Passivity refers to the tendency of some metals in corrosive, oxidizing environments to form surface films that render these metals more noble. Copper and aluminum have this capacity. Corrosion of the metal forms a tight surface film that blocks entrance of oxygen. This is very effective in stopping any more corrosion. Iron, on the other hand, forms a very loose film that actually encourages more corrosion.

Chromate solutions used in closed-water systems for heating and cooling are passivating solutions. Electrochemical action in the solution draws the chromium metal into a tight bond with the surface of metals to form a passive film on the surface. We are using the corrosion action of the base chromium to protect the more noble iron. The magnesium rod hung in the hot water tank is another example of using the corrosion of a base metal to protect a more noble metal. Compare Mg and Zn on the table to see how much potential is available for this. The serviceman who wants to see this action for himself can clip different metals to his milliammeter and dip them in a jar of water, salt water, or weak acid solutions. You will have a better understanding of the forces behind corrosion when you see the current flow. Don't use your microammeter unless you use very small metal pieces. Electromotive force increases with the area of the metals used.

Let's take a look at a corrosion process that most servicemen find very helpful. What happens inside a flashlight battery? A flashlight battery, as shown in Fig. 7-1, consists of a carbon electrode located in the center of a zinc-coated steel can. The outer surface of the carbon electrode is covered with carbon and manganese dioxide granules. The space between the carbon electrode and the zinc surface is filled with an electrolyte paste that is mostly ammonium chloride solution, NH_4Cl. An electrolyte is a solution that assists in the passage of current through it.

If you will connect a 1.5 volt incandescent bulb to the battery as shown, it will glow because the filament is heated by the passage of current through it. The current flow comes about in this manner. Zinc has a potential toward corrosion as shown in the table. Zinc is

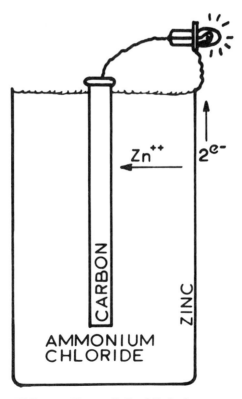

FIG. 7-1 Dry cell flashlight battery.

the largest area of metal available for corrosion in this device. As an atom of zinc corrodes it produces a Zn + + cation and two free electrons (2-). In the battery, the electrolytic solution provides water to aid in the corrosion process of zinc. The free electrons from the zinc travel up through the lamp and back to the carbon electrode. The cations attach themselves to water molecules and become hydrated zinc ions. These hydrated zinc ions travel readily through the electrolyte until they meet the manganese dioxide particles on the cathodic carbon electrode. Manganese dioxide will reduce and depolarize the hydrated zinc ions if electrons are available. The electrons given off by the corroding zinc that traveled through the filament, thus complete their circuit and end up in the chemical reduction at the cathode. When all of the zinc is used up by the corrosion process, the battery is dead.

Dry cells like this last only a few hours when they are connected to a closed circuit with a lamp. Put them on the shelf and they will last a long time with the circuit open. However, no dry cell will keep forever on the shelf. This is because local action currents can be set up in a battery. Minute impurities in the zinc. or differences in metal thickness, or even differences in temperature across the zinc surface, can produce cathodic areas that act like the carbon electrode. The areas around these cathodic points are called local action cells and the local action currents gradually destroy the zinc surface. Current flow is expended in wasted heat instead of useful light.

The two types of corrosion we have just seen in the flashlight battery have their counterparts in our service work. Connect copper and galvanized pipe and you have the potential for corrosion as shown in the tables. Pass water through the pipe and you have supplied an electrolytic solution. Any chemicals or salts that are in that solution

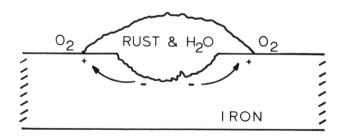

FIG. 7-2 Local action cell.

are going to make it more effective. The base metal in this coupling is going to be corroded by the type of corrosion that we know as electrolysis. The next time you see a connection leaking and find the base metal eaten away, you will know that it was caused by a closed circuit that allowed electron flow. Install a dielectric coupling between the two dissimilar metals to open the circuit and you will reduce the electrolysis.

Local action currents can function in a metal surface and local action cells can be built up on the surface of a metal, Fig. 7-2. This is because the impurities and/or alloys in the metals function to set up cathodic points in the metal that are electrically bonded to the remaining anodic areas.

Electrolytes are still necessary for local action currents, so as long as the metal is kept clean and dry, or painted, oiled, or in some way kept free of electrolytes, it will not corrode. The minute you expose metals to condenser water or boiler water, or a moist atmosphere, it will start to rust.

When iron rusts, the freed electrons travel through the metal to the cathode point. The hydrated iron ions, $Fe++$ travel through the electrolyte to the cathode. Salts, chlorides, and sulphates in the water provide the chemicals for reduction of the cations. The loose ferrous oxide and ferric oxide that is the result of this reduction process builds up over the local cell. Concentrated acid solutions build up in the cell and, if scale is present, a blister may form under which this local action cell is free to operate until it eats its way through the metal wall. This is the reason for the pitting so common on boiler tubes. Rust on metal surfaces may seem to be uniform, but when you wire brush it off, you find the underlying surface is rough and pitted. If you can count the number of pits on the surface, you can tell how many local action cells were involved in the corrosion process.

Controlling corrosion on metal surfaces that must be exposed to water solutions is done in a good many ways. Adding the yellow chromate passivating solution to the water in a boiler or cooling system heat exchanger is thought to form a diffusion barrier over the surface of the metal that changes the oxidation potential of the metal to that of a more noble state. Some passivating solutions also take up the dissolved oxygen in the water that is necessary to the reduction process. Phosphate flakes and crystals react to alter oxygen use.

The amount of treatment, or inhibitor, as it is often called, to be added to the water in a closed system is set by the chemist who analyzes the water. On a chromate solution, the right amount colors the water. A sample bottle of the right color is used to compare with samples that are periodically drawn from the system. Treatment is added until the system water is the same color. Treatments that do not color the water require use of a simple test set that is supplied with the chemicals. Maintaining the inhibitor at the proper level is very important to its effectiveness and the serviceman who is responsible for this treatment must follow the steps in the testing procedure and see that only the correct amounts of chemicals are added. Too much added to the system is often much more damaging to the system than no treatment at all.

I opened up an electric high-pressure steam press boiler in Davenport, Iowa, and had to use a hammer and long-handled chisel to remove excess chemical that had formed a solid mass on the interior surfaces. They burned out all of the heating elements because they had prevented water from contacting them. Phosphate or polyphosphate compounds added to a system not only form a surface coating to protect surfaces, but also help in keeping scale-forming solids in suspension in the water. The phosphate crystals or flakes in mesh bags hung in cooling towers are examples of this type of treatment. They do a very good job, but only if used in the proper amounts and if the tower basin is cleaned every month to remove dirt washed from the air. Failure to do this can result in a service problem that will be a real headache for the serviceman.

A large centrifugal system indicated, on the log sheets, that it was getting considerable condenser fouling by a wider and wider TD between water and refrigerant temperatures. I was sent out at the end of the season to clean this condenser. The man in charge of the equipment showed me the log sheets and commented that it had never been this bad in previous years. He could not understand it, because he had used a lot of treatment during the year. I looked at the cooling tower and found more bags of crystals hung in it than I had ever seen in a tower in my life. The bottom of the tower had about an inch of sludge and dirt in it.

I commented on the large amount of sludge and the operator said it was okay because it was below the level of the raised barrier around the leaving water sump. Now this is a mistake. The sludge and dirt you see in the tower basin is only the part that has come together in particles too large to remain suspended. The rest of the stuff is still in the water and is circulating with it. Phosphates increase the capacity of the water to hold scale in suspension. The more scale you have in the water, the more abrasive it is. The abrasive action of the water dissolved a lot more of the phosphate crystals. When the combination of scale and phosphate in the water reaches a certain level, it is going to plate out on the warmest surface. Either you keep the level down or it will plate out. There is no alternative.

We took off a condenser head and found the tubes were covered with a very hard, mud-colored layer that was as strong as a porcelain finish. It was fairly thick with a pebbled effect on the surface.

I had 50 pounds of dry acid and 5 pounds of soda ash neutralizer with me. I called the shop and asked that they send me 150 pounds more acid and 50 pounds more soda ash. I knew what I was up against. Phosphate is an inhibitor that protects the metal surface from corrosion. Coat the scale with phosphate and you protect it from the acid. Protection is not positive, it is just slowed down a great deal. The inhibited acid I was using would eventually eat through the glaze, but it might take a month and I could damage the steel parts of the condenser before I cleaned the tubes.

I set up the acid pump and started circulating water, then added enough acid to get a pH reading of 2. In thirty minutes I had a pH reading of 3. Another hour of circulation and I still had a pH reading of 3. I drained the system and refilled with fresh water. I added soda ash to get a pH reading of 9 and then dumped in a 10¢ can of lye. This caustic solution will dissolve the glaze better than acid and will not harm the metal. After two hours of circulation, I drained the system and refilled with fresh water.

This time I was able to use twice as much acid before I stopped losing the acid pH reading. I drained again and inspected the tubes. We still had a glaze but it was a lot thinner. I refilled with fresh water and left the system overnight. The next day I repeated the caustic circulation and then the acid circulation. Inspection after this showed the tubes to be clean.

If you ever run into this problem, don't try to do the job with one circulation of caustic and one circulation of acid. The first pass with acid takes out scale particles close to the surface. This leaves a rough surface with much more area. The caustic pass exposes more scale particles and makes the surface even rougher. As you alternate acid and caustic you eat through the glaze until it is honeycombed and finally breaks loose in large particles. You can see this action progress if you inspect the tubes between circulations. Now you know what can happen when you use too much of a treatment. Follow the manufacturer's recommendations. Do not exceed them. If more treatment would help, he would be the first to say so. Regardless of the type of treatment, keep the tower clean and maintain a constant bleed. There is a limit to how many solids can remain suspended in the water, and solids do build up because of evaporation.

EVAPORATIVE COOLING

Water will evaporate any time there is a difference in vapor pressure between the water and the atmosphere. Water, more than any other liquid, requires immense amounts of energy to achieve this change of state from liquid to gas. The huge amount of energy involved in this operation in large air masses is what causes some of the violent weather-storms that bring us rain and high winds.

The energy required for evaporation of water is taken from heat and pressure energy in the air and water. Our earliest forms of air conditioning made use of these facts. The evaporative cooler is a prime example of them. Air to cool a space is drawn through a housing that is lined with pads of excelsior, wood fibers that can soak up a lot of water both internally and on the surface. The water comes from a pan or sump and is continually recirculated over the pads. The purpose of the pads is to expose the greatest possible surface area of water to the air so that it can evaporate. As it evaporates, it takes heat energy from the remaining water and the air. The air temperature is lowered. Since we cannot destroy energy, what happens to it? It stays in the air in the form of water vapor. Energy has been changed from heat to higher relative humidity.

If the air was originally hot and dry, we can use this evaporative cooling to produce comfort conditions. Air that is cool and still below, say, 65 percent rh is acceptable for our comfort. We cannot use evaporative cooling if the air is originally hot and humid. The resulting cool air, at a high relative humidity, will not absorb perspiration and we will wind up with the cool, clammy feeling that results when our skin is covered with perspiration. Mildew, the growing of mold, shows up on shoes and clothing. Canned goods start losing their labels because the glue holding the labels loses its adhesive strength as it gets wet.

Air that is cooled with an evaporative cooler can only be used once. Every evaporative cooler must be a 100 percent fresh air system. Recirculation brings the rh up too high for comfort. The systems must have dry air to start with and this is why they are only recommended for use in dry climates. Failure to know and recognize the limitations of evaporative coolers resulted in their misapplication in a great many cases and earned them the derogatory name of *swamp coolers.*

Properly applied, evaporative coolers supply a lot of cooling for very little money. Every pound of water that evaporates requires about a thousand Btu's. Evaporating 12 pounds of water per hour is equal to a one-ton unit. That is twelve pints of water. There are many places even in humid climates where an evaporative cooler can be used. For instance: spot cooling for manufacturing operations or processes where high humidity is not a problem. Or, how about the air supply for an air-cooled condenser that is just not big enough when it hits 100 degrees outside? **Think** about it.

The big service problems with evaporative coolers are corrosion and scale. All the hardness that is in the water supply is left behind when the water evaporates. Once the pads become saturated with the salts left behind, the wood fibers no longer have the necessary wick action to draw the water through them. Surface area is drastically reduced and unit capacity drops by 50 or, in some cases, as much as 90 percent. The high mineral content of the water that is being recirculated makes it an excellent electrolyte and corrosion of metal surfaces under the scale is greatly accelerated. Scale and corrosion problems can be reduced if the water circulating system is equipped with a constant bleed.

CONSTANT BLEED

It used to be that, when I looked at a unit that cools water by evaporation and did not see the constant bleed, I accused the serviceman of failing to provide one. No more. Now, I ask him where he hid the constant bleed. Most units that cool water by evaporating part of it are water conservation units. They are used as a means of saving water. Those who do not know the reasons for a constant bleed raise quite a fuss when they see the open line and a steady stream of water running down the nearest drain. Let's make sure that we know the reason for it.

A 10-ton cooling tower is designed to reject 120,000 Btu of heat per hour to the atmosphere. To do this it must evaporate approximately 120 pounds of water per hour. That is more than 14 gallons per hour: fourteen gallons of water changing to pure water vapor and going off into the atmosphere; fourteen gallons of water at 5 grains hardness evaporating and leaving the 5 grains behind. Fourteen more gallons come in through the float valve to make up the loss by evaporation, but the hardness stays behind to mix with the fresh water.

If the cooling tower holds 140 gallons of water, all of the original is gone after 10 hours. The 140 gallons in the tower then contains 10 grains per gallon hardness. Another 10 hours and it is 15 grains; another 10 hours and 20 grains. How much mineral content can we suspend in the water before it combines with dirt washed from the air and precipitates out? How much can we stand before it plates out on the heat transfer tubes and cuts down heat transfer?

The maximum mineral content before trouble starts developing varies with the type of mineral. The best claim I have seen for tower treatments says it will hold up to 35 grains in suspension in average city water. Most systems will start scaling up at around 20 grains. How do we keep down the hardness buildup? The answer is to bleed off some of the water in the basin to get rid of the mineral accumulation. How much bleed-off? Bleed off as much as you evaporate. An acceptable **minimum** is one gallon per hour per ton. Combine this with a monthly cleaning of the cooling tower basin to get rid of the dirt washed from the air and you will go a long way in reducing scaling of heat transfer surfaces.

Hide the constant bleed if necessary, but make sure you check it every month. These bleeds handle such small amounts that they can easily be clogged. Take the valve handle off so people who find out about it do not turn it off. Most people do not know or understand the need for the constant bleed.

The actual amount of water evaporated varies from unit to unit. Some cooling comes from heat transfer to cooler air. The amount of water circulated per ton varies according to the type and design of tower. Let's look at some of them.

ATMOSPHERIC COOLING TOWERS

The first cooling towers I worked with were atmospheric-type towers, Fig. 7-3. An atmospheric-type tower has to have spray nozzles. The water must be sprayed through nozzles to produce as many fine droplets of water as possible. This is the only way that a large surface area of water can be exposed to the atmosphere for evaporation in a relatively small space. A nozzle that passes one gallon of water per minute at 60 psi pressure will produce more cooling of the water than a pond with a one acre surface area under the same weather conditions.

To produce the proper spray pattern for cooling requires three things: a clean nozzle, sufficient water quantity, and enough water pressure. Failure to provide all of these things will leave you with a tower that still sprays but does not have the cooling capacity to do the job. I have been called in many times to check out installations that could not keep the head pressure down when one look at the tower spray pattern told me what the trouble was. It still amazes me that servicemen can look at such a tower and fail to see the answer.

When an atmospheric tower is operating properly, the spray pattern is so fine, and so forceful, that the entire interior area of the tower is filled with a fog of these water droplets. You should be able to feel the airborne drops on your face when you stand on the down-wind side of the tower. Anytime you can look down through a tower and see large drops of water, even if there are enough to dimple the entire water surface in the basin, you are not producing rated cooling capacity. The fact that you can see the water in the basin clearly through the spray should tell you that something is wrong.

A spray nozzle is a precision device designed to take a certain quantity of water, flowing at a certain velocity, and start it rotating in a certain direction. We are imposing energy of a certain type on this water. As it is forced through the small opening in the nozzle, this energy is used to break the column of water into the fine droplets that allow it to present a large surface area to the atmosphere and completely fill the interior of the tower. A clean nozzle is a must. Anything that prevents the proper flow pattern wastes the energy used to pressurize the water. It results in turbulence just as scale in a

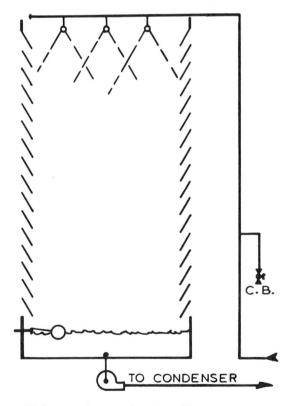

FIG. 7-3 *Atmospheric cooling tower.*

venturi or ice on a wing does.

Most atmospheric towers are designed to produce rated capacity with a wind velocity of 3 mph. With no wind movement, the air mass around the tower will quickly reach a state of equilibrium, as far as vapor pressures are concerned, and no further cooling will be possible. Higher wind velocities may or may not increase the cooling capacity. They will definitely increase water drift from the tower. I have seen towers with the upwind sides covered with burlap sheets to cut down drift. Some installations that have steady high winds have vertical baffle fences on the upwind sides to cut down drift. What helps during high wind velocities can be harmful during low wind velocities.

Each installation must be considered on its own merits. Location, nearby buildings, walls, **flues,** and exhaust openings must be taken into consideration. Any water spray system is also an air washing system. Whatever is washed from the air ends up in the water. What will be the effect of adding these substances to the water? **Think** about it!

A flue from a gas-fired hot water heater that is upwind of a cooling tower may allow flue gases to be washed from the air to mix with the tower water. CO_2 and SO_2 are common flue gases. Added to the tower water, they will make it acidic. In the proper amounts, they will help dissolve scale in the system and there are firms that utilize flue gases to control scale buildup in tower water. Amounts are carefully monitored and neutralized if the pH gets too low. A pH that is consistently too low will result in excessive corrosion of metals in the piping and condensers.

You can imagine what the results will be if greasy soot from a gas-fired hot water heater is drawn into the tower. And, what about the results if an exhaust vent from the kitchen range hood of a restaurant is also directed into the tower? For that matter, what if they are being drawn into an **air-cooled** condenser? An exhaust fan and a low flue from the hot water heater in the second building, caused the complete destruction of the aluminum fin to copper tube bond on a large air-cooled condenser in Wichita, Kansas, not too long ago.

Worn nozzles, scale in nozzles and piping, clogged strainers and pump impellors, all these things can cause troubles with atmospheric towers. The main thing is for the serviceman to be able to recognize the problem when he sees it, and then find out what caused the problem. Look for that tower filling water fog that shows that you are operating correctly. Don't cut down water pressure to cut down drift. You are only asking for trouble when you do this.

FORCED DRAFT TOWERS

Forced draft towers, Fig. 7-4, do not depend on wind currents for air movement. The required quantity of air is supplied by a fan or fans to produce the designed capacity. Spray nozzles are seldom used in forced draft towers. The required quantity of water is spread by water deck orifices or troughs to completely cover and saturate the internal fill. This fill is designed to have sufficient surface area to expose the water to the air for evaporation. Eliminator sections are designed to cut down drift. There are times and conditions when drift is still a problem from these towers. Strong winds in the same flow direction as the tower can cause excessive drift. Locating the tower to oppose the wind direction can cause problems. Propellor fans may not be able to buck the prevailing wind.

Induced draft is another term for forced draft. It is usually applied to larger towers. Large diameter, cast aluminum blade propellor fans, driven through gear reducers, move large amounts of air at slow speeds. Forced draft towers require less space than atmospheric towers and can be located with air ducted to and from them if necessary. The absence of fine spray nozzles reduces the attention needed to spray patterns, but the fill construction necessary for evaporation surface creates its own special problem. Birds find this fill construc-

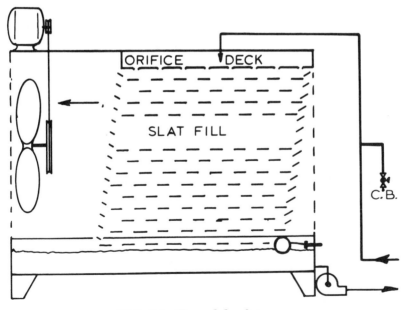

FIG. 7-4 Forced draft tower.

tion ideal for nest building and roosting space. The resulting debris and filth creates cleaning and sanitation problems. Most forced draft towers have, or should have, bird screens installed.

With the large quantities of air moving through a smaller space, these towers do a good job of washing the air and soon cover the bottom of the basin with dirt and sludge. Also, the smaller size of the towers in relation to the air volumes means a faster buildup of hardness in the water. Encrusted scale buildup on the fill material, Figs. 7-5 and 7-6, due to insufficient constant bleed, is very common on these towers. Corrosion of metal surfaces, and fans on pull-through units, is a major service item.

The cast aluminum airfoils of the larger units may be up to 12 feet long. Blade pitch is adjustable, as the blade is bolted to the hub by a center bolt. Pitch angle is usually set when the blades are bolted to the hub as the tower is assembled. Blade fastening bolts should be checked often to be sure that the pitch angle does not change and that the blade does not fly off. These fans may be 20 feet in diameter and damage, if one works loose, can be very costly.

FIG. 7-5 *Powdery tower scale produced by insufficient water treatment.*

Lubrication of the gear reducers on these large fans is very important. Because of their location in the towers, the oils used are a special industrial type with antioxidation chemicals and inhibitors added. Oil should be drained and replaced at the end of the cooling season, or annually, to get rid of accumulated moisture. If a customer objects to this, quote him a price for gears and bearings. He will agree that an oil change is cheaper.

The amount of water that must be circulated varies with the type of tower and system. If pump replacements are necessary, see if the gpm rating is on the old pump. If not, figure 3 gallons per minute per ton on towers without spray nozzles. Use 4 gpm per ton where you have spray nozzles. I have never seen an atmospheric tower that used less than this.

EVAPORATIVE CONDENSERS

Evaporative condensers, Fig. 7-7, combine the advantages of a tower and a condenser in one unit. They also place all of the combined service problems in one unit. Pay special attention to the constant bleed and tower cleaning. Scale removal from evaporative con-

FIG. 7-6 Granular tower scale produced by no water treatment.

densers, especially if they have finned tube condensers, is a lot harder than where you can confine the solution inside a tube. I once used eleven 13-gallon carboys of muriatic acid and a gallon of *Immurodine* inhibitor on a 75-ton Westinghouse evaporative condenser.

The practice of cycling the water pump instead of the fan on this installation was partly responsible for the excessive scale buildup on the tubes. The wiring was changed to cycle the fan instead of the pump, constant bleed was set up to waste 2 gpm, and an electronic water treater was installed on the supply water line. This was one of the very few electronic water treaters I've seen that worked. In four months, all the remaining scale inside the unit had disappeared.

Spray nozzles in evaporative condensers must be checked at regular intervals to insure that all of them are working. They must be patterned to cover the condenser surface completely. Evaporation is accomplished by covering the surface of the condenser metal with a film of water and then taking heat of evaporation from the metal.

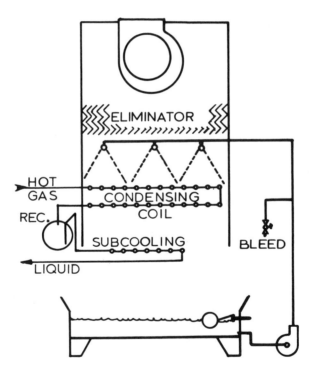

FIG. 7-7 Evaporative condenser.

Only a little air cooling is done by the water droplets below the condenser tubes in the air stream. Liquid subcooling lines are located below the tubes and may be found below the water level in some cases.

Squirrel-cage fans on evaporative condensers must be kept clean and free from rust. Air volumes are critical. Pull-through fans are especially susceptible to rust and it is not unusual to find eliminator plates badly corroded. I cannot overemphasize the need for cleaning and rust preventative maintenance on evaporative condensers. Immediately after fall shutdown is when most corrosion occurs. Don't wait too long to clean these units.

ALGAE

Algae and slime are problems that we encounter quite often in water used in heat transfer. Any system that is open, so that the water is exposed to the atmosphere, can and usually does develop algae and slime. This is because they develop from spores carried in the air that are washed from the air by the water.

Algae are rudimentary plants. Some of them are one cell plants, microscopic in size. They come in many shapes and are extremely important in the total life cycle. They provide food for fish, men and cattle. They possess chlorophyll, carry on photosynthesis, and thus are able to produce their own food. For all their importance to our economy, they are nothing but a headache when they start growing in our circulating water systems.

Algae block circulation of water when they plug screens and strainers and they reduce evaporation when they cover the surfaces in towers. Some of them feed on bird droppings and some use these nutrients for chemicals to produce large plants. These may be seen as the common green algae and less common red algae found in cooling towers. If the serviceman wants to learn more about algae, ask for information about plants under the family name, Thallophyta. It is a very interesting subject and your local library should carry the books.

Prevention of algae and slime in equipment is primarily a matter of sanitation and treatment. The problem varies widely with the location of the equipment, season of the year, type of water, and mineral content of the water. Treatment of the problem may be very simple or it may require a lot of work and chemical treatment. I know of some servicemen who place about fifty feet of scrap copper tubing in the tower basin and never have any more algae problems. You might try this, but don't be disappointed if it doesn't work. Water and air conditions have to be just right.

The two most common algaecides are copper sulfate, commonly known as blue vitriol, and the sodium hypochlorite solutions such as *Clorox* or *Purex*. Both of these solutions can cause damage to the equipment if not used carefully. If the system has an algae or slime problem that is small or only appears at intervals, these algaecides can be used as sterilizing agents. Clean the tower and piping thoroughly, using sal soda or TSP (trisodium phosphate) to get out any oily scum that may be coating the surfaces. Scrape and brush or sweep the basin to get out loose scale and dirt.

Refill with fresh water and then add the algaecide until you have just enough to detect the color change in the water: blue for the copper sulfate and yellow for the hypochlorite. Don't overdo it. A faint color indication is all that is necessary. Circulate this solution for at least two hours to do a good job of sterilizing all the surfaces, then drain and refill with fresh water. Systems that show light algae growth should have this done at the end of the season. This will kill the algae and wash it out so that dried spores will not be present at spring startup. Always flush and drain after this treatment. Corrosion problems will increase if you don't.

If the algae problems are too great to be handled in this manner, then you will have to set up a treatment program. There are several companies that supply algaecides and your refrigeration or heating wholesaler has them along with the literature on how to use them. Pick the product you are going to use. Get the literature and read it. Then read it again. Set up the program you decide to follow and carry it out all the way. With the large algae problems there is no such thing as a one-shot treatment. The treatment has to be followed continuously through the season. The type of treatment may have to be changed. After you get rid of one kind of algae in the system, another type may flourish because it no longer has to compete with the first kind.

You may have to clean and sterilize the tower between different algaecides. You will certainly have to clean the tower anytime you have a major kill of the algae. Dead algae plugs up strainers fast and can plate out on condenser tubes to form a leathery film that is resistant to acid cleaning. Never operate the refrigeration equipment while you are attempting a major algae cleanup.

All algaecides are toxic. Be very careful when using them. Read the safety instructions before using them. Some can only be used outdoors. Some can be used in air washers that condition room air. If you make the mistake of using an outdoor-only algaecide in an air washer, you will be held responsible for the consequences. **Think about it.**

WATER TREATMENT

Water treatment is required with almost all cooling towers and evaporative condensers. Almost every water supply has some minerals in it and the concentration of these minerals increases during the evaporation process. The constant bleed reduces the concentration but does not eliminate the need for treatment to reduce scaling and corrosion. The treatment procedure can operate along with the algaecide treatment. Suppliers of chemicals usually carry both scale and corrosion treatment chemicals, as well as algaecides. The two chemicals are compatible and can be combined into one operation. As with algaecides, pick your supplier, read the literature carefully, plan the treatment you will use, follow the treatment plan in full.

There is a wide variety of water treatments available to the serviceman. Most of those I am familiar with are the polyphosphate crystal type. These increase the ability of the water to hold minerals in suspension. To some extent, they raise the level at which mineral content plates out on hot metal surfaces. They do not take minerals out of the water. Only the constant bleed does this and practically all water treatment literature emphasizes this fact. The one point I feel is not emphasized enough in this literature is the need to keep the tower clean.

The polyphosphates that help keep minerals in suspension also help keep air-washed dirt in suspension in tower water. Take a sample of the cleanest tower water and allow it to stand in a glass overnight. You will be surprised to see how much dirt settles out. My experience has been that very few towers do not need cleaning at least once a month. Air pollution is not something new. It has always been with us. It is just lately that it has gotten the necessary publicity to make us all aware of the size of the problem. If you will stop and think about it, every tower, evaporative condenser, or air washer, is an aid to reducing air pollution. These pollutants, washed from the air, just add to our service problems.

Mesh bags of crystals in the water spray stay fairly clean. Screen-type crystal containers in the basin can easily be made ineffective by layers of dirt. They should be screened and washed on monthly inspections. Do this outside of the tower. The smaller particles of crystals can damage pump impellors, mechanical seals, and plug line strainers.

Liquid-type treatments are very effective. Flow rates are easier to control than crystal types. Do not try to change the flow rate from the calibrated tubes. If more treatment is needed, add another jug and tube. Check the pH of the water with test papers. Acidic water increases corrosion and also dissolving of phosphate crystals. When you find low pH readings, add pH neutralizer blocks. Don't dump in some baking soda. This is bicarbonate of soda and makes temporary hardness, the type that builds scale when heated.

Read all of the treatment literature, follow all of the treatment steps, do not exceed treatment rates, and you will do a good job of holding down scale and corrosion in the equipment. If you have problems that the literature does not cover, write to the treatment company. Most of them have a department set up to help you with analysis and special problems. Use this help and you will do a better service job.

ELECTRONIC WATER TREATERS

A fairly recent development is the electronic water treater. I have seen many of these installed but very few of them that really worked and continued to work long enough to justify their cost. The theory behind them is basically sound. Scale buildup and corrosion are both electrochemical actions. Electron loss from the elements results in ionization of the elements. This is the action shown in the oxidation potential tables. All electronic treaters I have seen advertised operate on the theory of ionization of the water, and the minerals in it, before it has a chance to build up scale. The free electrons released in this process flow to the interior surfaces and reverse the corrosive electron flow from these surfaces.

The fact that the theory is sound, but the results are so poor, has caused me to do a lot of study and work in this field. I believe there are two reasons for the poor performance of electronic water treaters. The first is that any attempt to ionize is an electrochemical operation. The unit is designed to subject the water and the minerals in it to electron flow. Whether it is a small unit with a magnet in it, or a large, expensive unit with a power pack and high voltage direct current electrodes, the ultimate idea is electron bombardment. The units are self-destructive. Scale and corrosion are concentrated in the unit and this makes them the first point of destruction.

The second reason is the small amount of current flow that is available from these units. Even the power packs, that supply potentials as high as 400 volts, still deliver only microamperes of current. This is not enough to counter the current flow in scale and corrosion actions in the system. If you experimented with your ammeter and strips of metal in electrolyte solutions, as I mentioned earlier, you found out that you would get current flows in milliamperes from small size strips. I have placed my ammeter across dielectric couplings in pipe lines and registered current flow in amperes. The quantity and type of metal in these systems is what governs the amount of current. The cost of electronic treaters capable of delivering the necessary current and resisting destruction, would seem to be out of proportion to the benefits available from them.

WATER SOFTENERS

I have previously stated that I do not like to use water softeners on cooling or heating water systems and I should explain why. Water softeners are primarily used to get better cleaning action from soaps and detergents. Calcium and magnesium are minerals in hard water that react with soaps to form curds. These curds float on the water and can be redeposited on the clothes, trapping dirt with them. Hard water requires more soap to clean clothes and help wash out the curds.

Water softeners in common use today do not reduce the total mineral content of the water. They replace minerals in the water with minerals that are compatible with soaps, that do not produce dirt trapping curds. Water thus softened will do a better cleaning job and require less soap in the process.

Even though the total mineral content of the water has not been changed, this *soft* water will test out in a laboratory as having a *hardness* of 1 or 2 grains where it originally tested as having 4 or 5 grains of hardness. This is possible because it is difficult to identify the exact composition of hardness minerals. Standard testing procedures lump calcium and magnesium bicarbonates, carbonates and sulphates together and report the total amount of these compounds as if they were only carbonates. Five-grain water contains calcium and magnesium compounds that are equivalent to 5 grains of carbonates of these metals per gallon. Replace some of these minerals with sodium, which is not classified as *hard*, and you will get the lower 1 or 2 grain *hardness* reading.

I object to *soft* water in cooling or heating systems because water that is high in sodium is a much better electrolyte than water that has not been *softened*. Prove it to yourself. Take a sample of water before it goes through a water softener and another sample of the water that has been softened. Suspend a large nail in each sample and set them on the shelf. See which sample rusts the nail the most, the fastest.

There is no question about the benefits that soft water brings to the laundry, kitchen, bathroom, beauty parlor. But one of the costs of this benefit is the increased amount of corrosion in the equipment and piping carrying this water. Corrosion is magnified as this water is exposed to the atmosphere. Raising the mineral content by using this softened water in evaporative units, creates strong brines and rapid corrosion. I have seen several coil air-washing systems that were originally designed and installed with water softeners for the make-up water supply. Without exception, the operators have taken the softeners out of the system after the first season. They would rather have the scale problem than the more expensive corrosion problem.

Most of the water softeners used today are the *zeolite* type. A zeolite is a crystalline mineral in the hydrous aluminum silicate group.

One of the properties of some zeolites is that the crystalline structure acts as a sieve. Small molecules pass through, large ones are trapped in the crystalline structure. The molecular sieve drier we install in the refrigerant liquid line is an example of this. Some zeolites pass water molecules through. While the water molecules are in transit, they have the ability to exchange ions with the water. Sodium zeolites are very good at doing this. A sodium zeolite that has exchanged all of its sodium ions for calcium and magnesium ions, can have its action reversed by passing a strong salt solution through it and exchanging sodium ions for the calcium and magnesium ions. This is the regeneration process of the water softener.

The water softener has its place in our world today and is certainly beneficial when properly used. The serviceman should know its advantages and its disadvantages. This will enable him to put these water softeners to their proper use.

8

Air Circulation

The drawing, Fig. 8-1, represents a 3-ton condensing unit I worked on in 1970. It was on the roof of a shopping center and located just 6 feet north of the south wall. This wall was designed to hide the units on the roof from view of customers in the parking lot.

Air from the unit discharged south into the wall and was deflected up. The prevailing wind was from the south and as it came over the wall, it began to turbulate. I noticed this because it was carrying smoke from a trash fire about half a block away. The combined effect was to pick up the discharge air and recirculate it back through the condenser. The unit was going off on internal protector and staying off for several hours before resetting itself.

The suction line was not insulated after it came through the roof sleeve and did not seem too cool. A check with a thermometer showed suction line temperature to be varying between 72 degrees and 87 degrees. Because this unit was a burnout replacement, we cleaned

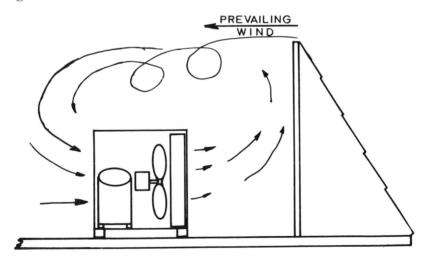

FIG. 8-1 Recirculation of condensing unit discharge.

the valve, installed a new liquid line drier, set superheat to 8 degrees, installed a discharge duct that turned the discharge up and ended six inches higher than the wall, and insulated the suction line. It solved the problem.

If a unit takes air in on the sides and discharges straight up, you should not have trouble with it. An exception would be a low wall close to the unit on the upwind side. This could possibly cause a problem due to turbulation. A unit that takes air in on one side and discharges on the other side should be installed so that it discharges in the direction to follow the prevailing wind. A propeller fan does not have enough power to buck a 5 to 10 mph wind.

These are just common sense rules that everyone would seem to be familiar with. If you have driven on the highways in flat country, you may have come into a small town with a grain elevator close to the highway and seen the sign *Watch for Wind Currents.* You may be holding the wheel to the left to compensate for the wind and when you hit the turbulence of wind flowing around the elevator, the car suddenly veers to the left. This is air flowing back in a vortex to fill the space on the downwind side of the obstruction. The combination of the prevailing wind and obstructions around an air-cooled condensing unit can do tricks with the unit's operation. There are no set rules in a case like this. You just have to use your head and figure out cause and effect for yourself.

The drawing of the underground condensing units, Fig. 8-2, shows a problem I ran into a few years back. The small building is about the size of a two-car garage and 2½ walls were storm louvers with 2-in. filters on the inside. This was the entrance and air intake room for an

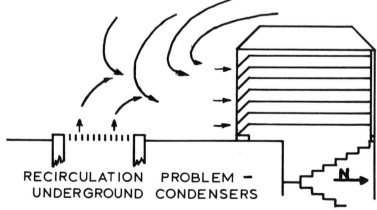

RECIRCULATION PROBLEM –
UNDERGROUND CONDENSERS

FIG. 8-2

underground building that had four 60-ton, air-cooled condensing units for the air conditioning. Air from the condensers was discharged at ground level sixteen feet south of the building. Prevailing winds in this area were from the southwest so, in theory, this was an acceptable installation.

In actual operation, there were enough winds from due south to seriously affect the operation of the equipment. I was working on this installation for over a month and frequently checked thermometers I had hung at several locations. The high condensing pressure alarm came on one day and I found that air was entering the condensing fans at 102 degrees. I went topside to see what had happened, because it had been a nice cool morning when I came to work. I found it was still a nice cool day. There was a breeze from the north and air temperature was 72 degrees. The wind from the north was turbulating over the top of the building and recirculating air from the discharge into the south wall entrance. I recommended a wall around the discharge opening to bring it above the building height as a solution to the problems. This was a fairly large opening, as the condensing units required a total of 151,000 cfm of air. Whether the unit is large or small, air problems can still cause trouble.

CIRCULATION

As long as we are talking about air, let's get into something that is one of the biggest problems in air conditioning. Circulation is one of the seven functions of air conditioning. It means the total amount of air that is circulated through the fan unit and in turn the conditioned area. It is very important from two points. The evaporator is designed to remove a certain amount of heat from the air passing through the coil. Some of this is removed as sensible heat and the temperature of the air is lowered. Some of it is removed as latent heat and the air is dehumidified.

The engineer who designs the system matches the capacity of the cooling coil and the condensing unit to the amount of air and the nominal amount of moisture in it to produce the correct amount of air conditioning for summer use. If this is an all-year system, he also matches the amount of air with the heat exchanger for winter use. This may be a coil supplying heat from steam or hot water, or it may be a gas or oil-fired furnace, or it may be electric resistance heat, or it may be the cooling coil switched to a condensing coil by means of a reversing valve.

Whatever the heat source, the volume of air required for the conditioned area is a prime factor in selecting the cooling and heating equipment that conditions this air. The volume of air circulated through the equipment and the heat energy added to or taken away from the air acts upon the equipment that we are servicing. The volume of air that is circulated in the conditioned area determines the comfort conditions in this area and, if the customer finds the comfort conditions are not satisfactory, he calls for the serviceman.

The serviceman who is going to do a good job on any air conditioning system must know about circulation. Let's see what it is and how it is determined. We will use the term cfm to define volume of circulation. Cfm stands for Cubic Feet per Minute.

The engineer who designs an air conditioning system first measures the *container* of the space to be conditioned: floor area, ceiling area, roof area, wall area, glass area and door area. These dimensions define the sizes of surfaces through which heat energy either enters or leaves the area to be conditioned. He notes the type of construction, amount of insulation, direction it faces, amount of sunlight exposure, wind exposure, shading, all things that can affect the amount of energy transmitted through the surfaces.

The engineer has tables that give the coefficients of heat transfer through the various types of construction in Btu per hour per degree temperature difference. This is called the U factor. He also has tables that give the outdoor design conditions for the geographical area. These design conditions tell him the average high summer temperatures and humidity and the average low winter temperatures and humidity. The difference between inside and outside design temperatures is the TD. Putting all this information together, area x U factor x TD gives the heat leakage through the surfaces in Btu per hour.

The next step is to total up the internal heat load. Heat from lights, appliances, and machinery are added for each room or conditioned area. Then heat from people, how many people, how active they are. Activity makes a difference. It takes more energy to work at a typewriter and adding machine than it does to sit and watch television and this difference in activity is charted as to the energy output of the individuals.

Next, the engineer will total up the amount of outside air that infiltrates into the building through doors and windows, and how much ventilation is required. Ventilation is one of the seven functions of air conditioning and, as such, must be a part of every heat load calculation. Even if you do not provide a separate ventilation inlet, you are going to have some ventilation infiltrating into the building. Every exhaust fan in a toilet, over a cooking range, exhausts air from the building and increases the infiltration through doors and window crackage. If a gas or oil-fired heating unit is installed in the building, it has to have combustion air and this air also contributes to air infiltration unless special provision is made for it.

The amount of infiltrated and ventilation air is totaled and the Btu load for sensible and latent cooling of this air is figured. Putting all these figures together, the engineer comes up with a total of the sensible and latent heat load for each individual area or room, and for the building as a whole.

Cfm required for each room and for the entire system is figured on a basis of Btu's per cfm. The figure varies, but an acceptable average figure is 30. If a room has a heat load of 3,000 Btu, then 3,000/30=100 cfm. If the grand total heat load is 36,000 Btu, then the system cfm is 1,200. Using these figures, the engineer can select the size of the equipment needed and design the ductwork necessary to bring the proper cfm to each room. If the engineering has been properly and carefully done, the installed equipment will do a satisfactory job of air conditioning.

The serviceman is going to run into service calls where he checks the equipment and it will seem that the machinery is working perfectly, but the customer still says it is not doing the job. Let me tell you about one such call.

This was the engineering section of a manufacturing firm. The air conditioning system was a 6-ton horizontal package unit with a water-cooled condenser and a forced draft cooling tower. The complaint was not new. It had happened the year before in periods of hot weather and several servicemen had been on these calls. I had checked it out thoroughly the summer before and reported on my call sheet that the equipment was doing all it was designed to do and it looked to me as if the equipment was not big enough for the load when summer weather changed to **hot** summer.

Evidently this comment had been read by our firm's president, who was being called by the customer, and he called me in to talk about it. He asked me what I based my comment on. I replied that I had made a complete check of all parts of the system and found the system to be operating at temperatures, pressures, air, and water quantities that indicated it was doing all it could. We went over these items that I had recorded on my call sheet and he agreed with me that the unit seemed to be delivering 100 percent.

"Where do we go from here?" he asked. "The president of this firm is my brother-in-law. We designed this system and sold it to him and I am not going to get him off my back until it works."

There is only one thing to do in a situation like this. I made a survey of the office and completed the heat load figures for it. Total load was 10.6 tons.

Replacing the present unit would have been quite a job. The biggest problem was changing the ductwork. There was no way to get the increased cfm needed through the old ducts. The physical layout of the office was such that I could locate package window units in a wall at one end and direct the air flow to mix well with the present system. I made this recommendation on my survey form and submitted it to the boss. Material and estimated labor costs were included. Our installation department received orders the next day to start installing the two 2-ton packages.

The job came up for review at the next monthly meeting of the executives and I was called in to explain why my figures called for ten tons when the original survey called for six tons. To me, the answer was simple. A comparison of the two surveys showed that the room dimensions had not changed, but there were a lot more people, double the number of lights, and electric typewriters, adding machines, calculators, and even a 40-cup coffee brewer on my survey that had not been on the original survey. The business of the manufacturing firm was growing rapidly and the engineering office had grown with it. The air conditioning installation had to grow too. I don't know how the boss and his brother-in-law came out on the money angle, but I do know that we stopped getting service calls saying the system was not cooling the office.

This was a case where a serviceman was able to say that the equipment was operating correctly, and prove his point that more equipment was needed, by a knowledge of design engineering. Let's look at another example.

This was a ranch-style, four-bedroom suburban home, well built, well insulated, shaded by large oak trees. I was trying to sell them a central air conditioning system. The heating system was an oil-fired 140,000 Btu Lennox furnace with aluminum ductwork. Registers were Honeywell high wall and return air grilles went to panned joists dumping into a return air duct that was parallel to the supply duct.

I was using a Frigidaire survey form that was designed to figure room-by-room cooling loads, heating loads, duct sizes and register sizes. I completed my survey and found that the installed ductwork was too small for air conditioning and undersized for a satisfactory heating job. I brought this up when I was talking to the homeowner and justifying my price. He said that he had never been able to keep the bedrooms warm enough in really cold weather and the basement was too cold for the family to use the workshop or for his wife to raise her African violets. Oil consumption was averaging 1600 gallons per year and the furnace required vacuum cleaning for soot removal every year.

I sold the job and installed a 3-ton Frigidaire fan, coil, and compressor unit next to the furnace with a Kramer Trenton air-cooled condenser and receiver outside. I was able to buy matching registers and installed them in the rooms my survey showed needed them. I removed all the joist pans and let the return air dump into the basement. The return air duct was disconnected from the furnace and tied into the supply air plenum. Supply runs were moved to balance flow in the two main supply ducts.

Two large plate-type humidifiers were installed on a mount over the furnace return air intake opening and a filter frame to cover the humidifiers. A pane of glass was removed from a window next to the furnace and a 12 x 18 duct with a 2 in. filter in it was run from the window to the burner housing of the furnace. Two dampers were installed to switch air flow from the furnace to the air conditioner when they wanted to change over.

What I had done was to add to the circulation capacity of the heating system, make the basement a return air plenum, and provide a positive source of combustion air to the oil burner. I set the furnace fan switch down from 140 degrees to 105 degrees.

The customer was very happy with the air conditioning system, every room was comfortable in the summer, and he no longer had a damp basement. With the thermostat set at 72 degrees in the winter, the bedrooms were 72 degrees even when it was below zero outside. What made him really happy was that his oil consumption for the next five winters was between 900 and 1,000 gallons each winter and the sooting problem almost disappeared. Correcting air problems was the answer in this case.

Another service call was on a three-story telephone exchange building. Some of the switching equipment is stepping style and requires a fairly high humidity for correct operation. The complaint was lack of circulation. The plant maintenance crew could not figure out why it had suddenly dropped off. All equipment seemed to be operating correctly and they asked if we could locate any concealed fire dampers that might have closed or ductwork insulation that might be causing the problem by coming loose from the inside of the ducts.

This was a hot deck, cold deck system with the built-up fan, filter, and coil assembly in the basement. Return air dampers and 100 percent outside air dampers were at the second floor level and the exhaust air damper was at the third floor ceiling with exhaust through the roof. According to the gauges on the pneumatic control board in the basement, all dampers were set correctly and we were using 20 percent fresh air. I inspected all these dampers to make sure they were actually operating as the gauges indicated. I found the outside air damper open, return air damper closed and exhaust damper closed. It was like trying to blow air into a pop bottle. When the air pressure in the building built up to the fan cutoff level, circulation stopped. We set the dampers manually for correct operation and had to have the pneumatic control people come in and clean and recalibrate the instruments so that they would operate automatically.

This would seem like a simple problem, but I have been called in on similar problems several times after servicemen have failed to spot it. I recall a drugstore in Missouri and a department store in Minnesota. Exhaust air was taken care of by aluminum shutter-type exhaust openings. Wind gusts move these lightweight shutters and sometimes cause annoying air movements through them. In both cases, the openings had been covered over solidly by store personnel. The control system had no connection to the exhaust shutters, but servicemen seeing the 100 percent outside air intake damper should have checked for an exhaust opening. The fact that they did not indicates to me that they were not knowledgeable about circulation on such systems or were **not thinking** when they made the service calls. In both cases, we took out the metal shuttered dampers and installed neoprene coated fabric shutter openings. They are quiet in operation and much less affected by wind gusts.

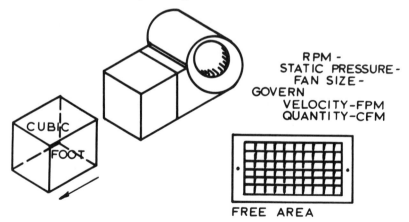

FIG. 8-3   Free area equals 60 to 85 percent of grille area.

MEASURING CIRCULATION

Since circulation is so important in air conditioning systems, the serviceman must know how it is measured and what instruments are used for this purpose. Circulation is expressed in cubic feet of air per minute; quantity, distance and time are the ingredients. If air leaves a duct one foot square at a velocity of 100 fpm, then we are moving 100 cubic feet of air per minute. If the velocity remains 100 fpm but the opening is two feet by three feet, then we are moving 600 cfm. For the serviceman, cfm becomes velocity times area, because velocity is the combination of distance and time. Area of the opening is what determines the quantity.

Area becomes complicated when the serviceman works with grilles and registers. A grille is a perforated or formed opening for the passage of air. A register is a grille with an adjustable damper behind it. A 12 in. by 12 in. grille does not have a square foot of **free area** for air passage. Free area, Fig. 8-3, may be anywhere from 60 to 85 percent of the grille size, depending on the design. Manufacturer's catalogs spell out the free area for the different styles and types of grilles and registers.

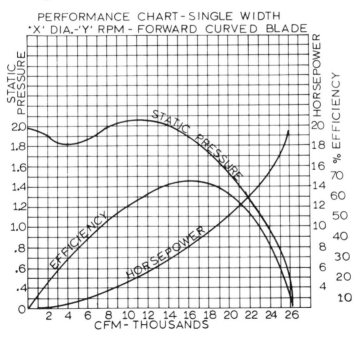

FIG. 8-4 Typical fan curve chart.

The amount of air that a fan will move depends on two things: speed and pressure differential across the fan. Fan manufacturers have tables or fan curve charts, Fig. 8-4, that give fan capacities in cfm at different rpm's and static pressures.

The instrument that measures rpm is the tachometer. There is no such thing as a cheap tachometer. I have used some very good ones, but could never justify the high cost of a good tach. A revolution counter, such as the one in Fig. 8-5, is not expensive and, combined with a watch with a sweep second hand, gives the same results as a tachometer. Learn how to use it and every time you use it take three readings and compare them. If the readings are close, then you are using the counter correctly.

Static pressures are measured in inches of water with manometers. Two kinds are available. The U-tube manometer in Fig. 8-6 has *plastic* tubes and comes in 7 inch and 15 inch ranges. These instruments use a special fluid that allows the overall size to be cut down to almost half the size that would be necessary if water were used. Scales are graduated in tenths of an inch.

Inclined tube manometers, such as the one in Fig. 8-7, are used to measure static pressures in hundredths of an inch. An incline that rises one-half inch as it travels horizontally five inches, plus a fluid with a specific gravity that is less than water, allows the wide spacing necessary for the small measurements. The one shown has a minus .05 to plus .50 range. Inclined tube manometers are used to detect the pressure drop across air filters. The dirtier the filter, the more pressure drop.

The 16-inch manometer shown in Fig. 8-8 uses distilled water and is a very old service tool that was used to set gas pressure regulating

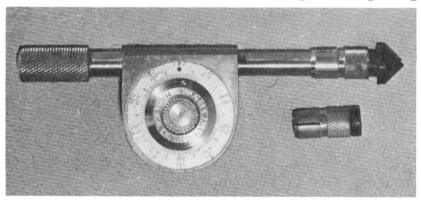

FIG. 8-5 Revolution counter used for measuring rpm.

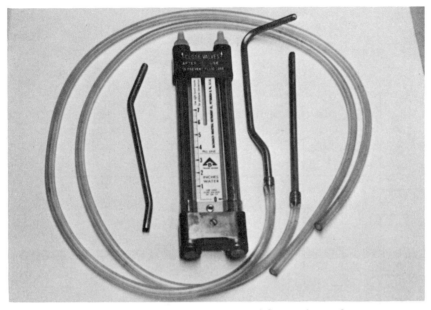

FIG. 8-6 *U-tube manometer with sensing tubes.*

FIG. 8-7 *Inclined tube manometer.*

COMPARISON CHART

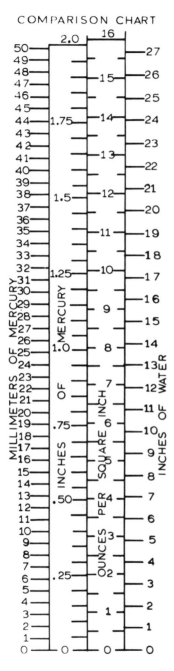

Chart 8-1

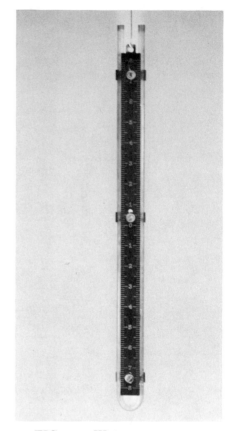

FIG. 8-8   Water manometer.

valves on Servel Electrolux gas refrigerators. A tenth of an inch pressure difference in the gas supply made a world of difference in the operation of these absorption units. The new compact manometers, with valves for closing off the U tubes, are a much handier and more accurate tool for today's serviceman.

A manometer can be used to measure vacuum or pressure depending on which tube you hook the hose to. You can leave one tube open to the atmosphere and use the hose on the other tube to read pressures at different places on the air unit or ductwork. The difference between the readings will indicate pressure drops across items such as fans, coils, filters, Fig. 8-9. These comparisons are accurate only as long as the open tube is exposed to the same atmospheric pressure. Accurate readings of pressure drop across an item or group of items can be made with two hoses connected to the tubes. If long hoses are necessary, allow time for the pressure in the hoses to equalize with the chamber being measured. Five minutes for ten feet of quarter-inch hose or tubing is enough.

If you are using the manometer to set valves to a specified delivery pressure, take a good look at the instructions and see if they also specify the atmospheric pressure. If they do **not** call for a specific atmospheric pressure, do not worry about it. The other side of the valve diaphragm is exposed to the same lower atmospheric pressure and your setting will be accurate. If they do, you will have to refer to a barometer or call the Weather Bureau to find out what it is. If the specs call for atmospheric pressure of 29.92 inches of mercury and the barometer reads 29.62, your manometer will not indicate correctly. A .30 inch (of mercury) lower pressure means the open tube of your water manometer will read 4 inches high, Chart 8-1. **Think** and use the tables to find the necessary conversions.

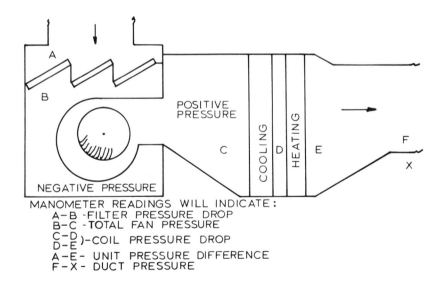

FIG. 8-9

Mercury tube manometers are also available and are used on equipment where pressure and vacuum readings are beyond the normal range of water manometers. Absorption systems use them and the McLeod vacuum gauge is designed for a situation where only mercury can be used. Mercury does not evaporate under low pressures at normal temperatures.

A manometer is a tool that the serviceman will find extremely useful in his work. As an example, a reading of the pressure drop across a cooling coil before and after it has been cleaned tells you a lot. Comparison of pressure drop across a clean, newly installed coil and the same coil after a year of use could tell you a lot. **Think** now, this would only be valid if you were putting the same amount of air through the coil. You would have to have fan speeds and static pressures across the fan before and at the present time to prove this. If you found differences here, that would tell you something too, wouldn't it?

Be careful when you take manometer readings. The position of your pressure tube in the chamber in respect to the air flow makes a great difference in the reading. Air that is moving in a direction that will force it into the tube will indicate static pressure plus velocity pressure, Fig 8-10. Use a curved tube on your manometer sometime and rotate it so that it gets velocity pressure. Compare it with readings with the tube at right angles to the air flow and with the tube turned in the same direction as air flow. You will learn a lot about using your instrument by these experiments.

Velocity flow is how an aircraft air speed indicator works. To read air velocity with a manometer, static pressure, temperature, and humidity must be known, and extensive tables are necessary to con-

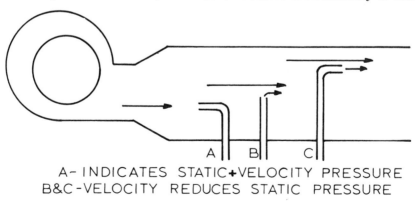

A- INDICATES STATIC+VELOCITY PRESSURE
B&C -VELOCITY REDUCES STATIC PRESSURE

FIG. 8-10

vert this information into accurate data. There are other instruments available for this purpose that are much easier for us to use.

The serviceman will get more accurate static pressure readings with his manometer if he uses a sensing tube on the end of the flexible tubing, Fig. 8-11. Use a piece of tubing with a round closed end. Soldering a small round head screw in the tubing end will do the job. Drill holes as shown in Fig. 8-12. An offset in the tube enables you to judge the position of the tube in the ductwork. Some of my sensing tubes are shown in the photo of the manometer.

There are many instruments for measuring air velocity. The Weather Bureau uses an anemometer, four half-circle cups rotated by the wind. Speed of rotation is converted, usually electrically, to the dial indicator. Wind speed is represented in miles per hour. The air conditioning engineer's anemometer is a multi-bladed fan in a case that looks like an alarm clock and is about the same size. The fan is connected by gears to dials that record the fan revolutions in terms of cubic feet of air. The instrument is moved slowly over the face of the air discharge area for five minutes. Dividing the air quantity recorded by 5 gives the velocity. Air velocity is not always uniform over

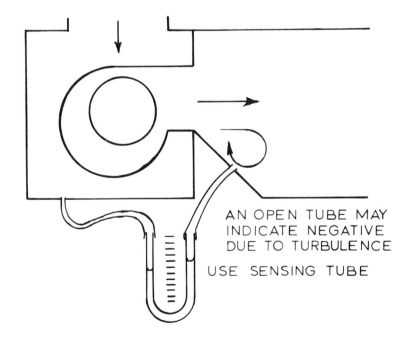

AN OPEN TUBE MAY
INDICATE NEGATIVE
DUE TO TURBULENCE
USE SENSING TUBE

FIG. 8-11

the entire face of a grille or register. The movement, to cover the entire face area, and the extra time allows you to get a more accurate measurement.

The Alnor Velometer is a very accurate and expensive instrument. A heated thermocouple passes current according to the temperature of the probe. Velocity of air over the probe determines probe temperature. With the accessories that come with it, the Alnor Velometer can be used to read pressures, velocities, and temperatures in almost any part of an air conditioning system. The serviceman would have to be working on many large ductwork systems to justify the purchase of this instrument.

The average serviceman will get good results with velocity meters such as the Bacharach Florite or Floret, or the Alnor Velometer Jr., shown in Fig. 8-13. These are accurate enough for the serviceman to do a very good service job on any systems that are within the instrument's range. Air passing through an orifice deflects a foil covered vane to indicate velocity in fpm. Large and small orifice allow use of low and high range scales.

When you use a velocity meter you must keep in mind the fact that you are dealing with fluid energy: air in motion, air that is trying to go from here to there. And, after using a velocity meter and seeing some of the readings you get, you will find that air in motion follows the old saying that a straight line is the shortest distance between two points. The fact that air is compressible allows it to vary its velocity over the face of a coil, or filter rack, or intake grille, or discharge register. It will even follow this behavior pattern out into the room after it leaves the register.

When you are measuring air velocity in or out of any opening,

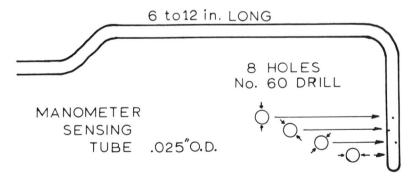

FIG. 8-12

always break the area down into smaller areas and take a measurement in each of these areas. Add up all these readings and divide the total by the number of readings. Even a small register, 4 by 8 inches, can surprise you by the amount of variation between readings taken at four or eight places in front of it. Read the air velocity at 27 locations in front of a 36 by 96 cooling coil and you will know what is meant by unequal air distribution.

Were you standing in front of the coil when you took these readings? Then you were blocking the air flow pattern. Hang your meter from a wire or mount it on a small pipe tripod while you stand off to the side when you take the reading. You not only have to **think** about what you are doing but also **how** you are doing it.

After a serviceman has used a velocity meter for a year or two, he will have a much better understanding and appreciation of proper duct design, proper transition of duct dimensions, and the need for turning vanes and air mixing chambers.

One of the most common and sometimes most difficult of service calls is the complaint that some rooms are being undercooled or underheated. The usual answer is to get more air into the room. While

FIG. 8-13 Alnor Velometer Jr.

this is true, it is sometimes easier said than done. Can you get more air through the register? Will the duct design allow you to do this? Will this take air from some other room and simply shift the problem to that room?

If you were able to get more air into the room and it solved the problem, that is fine. If you get called back because it did not solve the problem, don't waste time on individual room air quantities. These could be symptoms of the real problem. How much total cfm are we moving and how much should we move? This is where the serviceman has to know the facts of life about this particular air conditioning system.

Was the system properly engineered and designed before it was sold? If so, a duct layout and air volume sheet should be available. Working from this, the serviceman can check the fan for speed, static pressure, and air velocity to see that it is actually handling the required amount of air. After you have established that you are moving the design cfm, or made the necessary repairs to obtain design cfm, you can proceed to set the outlets to the design cfm of each.

This balancing of a system is not easy. My own method is to open all the outlet dampers or registers wide and then start in close to the fan and work out to the farthest outlet. Repeat this sequence as often as necessary until you obtain the proper settings. Every time you pinch down a damper you will be forcing a little more air through the dampers you have already set. It takes patience, time, **thinking,** and a knowledge of the duct system and how it works to do the job of air balancing. There are firms that specialize in air balancing, but on the residential and small commercial jobs the serviceman is often stuck with this chore. He must be able to do the job if he is going to call himself a serviceman.

If the serviceman balances the system and finds that it still will not do the job it was supposed to, then, and only then, can he go back to the engineering department and say, "Fellows, it's your headache." Don't let them bluff you out at this point. If you have done your job correctly and have all the readings down on paper to show that you did, you can pass the buck back to them. After all, engineers put their pants on one leg at a time just like servicemen and they can make mistakes, just like servicemen.

All too often the serviceman finds he has circulation problems on installations that have no engineering background that he can refer back to. It may be a central air conditioning system that was added on to an existing heating system. It may be an all-year system that was installed where installed price took precedence over engineering standards. Whatever the reason, when a serviceman runs into a call of this nature, he has to have some sort of reference point to work from. All service work calls for logical thinking and logical thinking demands a reference point from which to start. The starting point in a call of this nature is: what is the amount of heat energy that must be moved to accomplish the desired results? The serviceman, who has no engineering department to call in for help, must be able to make his own heat load surveys in order to establish the job requirements.

I realize that this statement is going to bring down the wrath of the HVAC engineering profession on my head. There are going to be claims that half-baked heat load surveys by servicemen will only make the situation worse. And these claims will be correct if the serviceman does not learn how to do the job right.

The fact remains that too many of the residential and small commercial air conditioning systems installed today are either underengineered or not engineered at all, due to the competitive nature of our economy. The serviceman who checks out the operation of the installed equipment, and finds it is doing all it is designed to do, is in a spot when the customer asks why it will not condition the space to comfortable temperatures. He cannot earn and collect his money unless he can come up with the solution to the problem. When he does come up with a solution that cures the problem and satisfies the customer, he gains another of the steady customers that are so necessary for a profitable service business.

A basic, usable knowledge of system design is a must for the majority of men servicing heating, cooling, or all-year systems. The men servicing commercial systems that are designed for precise temperature, humidity, ventilation, and air cleanliness control may not understand how the system is supposed to do this unless they have this basic knowledge.

I am not going to attempt to cover heat loads and system design. There are many books on this subject. Unless you are very lucky and extremely knowledgeable, you will not be able to learn this subject just by reading a book. It takes guided study and lots of questions along with the correct answers to your questions to learn the subject. The major manufacturers of equipment and the industry associations have books and study courses on the subject. Talk to these people, find out what is available to you. Decide what course you want to take and then take it. Whether it is a correspondence course, or regularly scheduled classes, keep on schedule, don't miss any classes and don't take too much time between mailing in the lessons. If any item, statement or terminology used is not clear to you, ask about it. There is nothing in this subject that is beyond the comprehension of the serviceman.

The knowledge that a serviceman gains from a course such as this is extremely helpful in understanding some of the happenings in service work. The reason why a heating system does a better job and uses less fuel when more air is circulated over the heat exchanger will become clear to you. The cause of complaints that, "It's cool in here but we keep on perspiring," are answered with a knowledge of design engineering. No book can **tell** you the answers to all of the service problems you may run into. **Knowledge** and **thinking** on your part is the only way to find the answers.

Before leaving the subject of air, I want to emphasize one phase of air circulation that is very important. It affects every all-year system that has a combustion chamber for burning fuels. Failure to consider the need for combustion air could cause the serviceman to fail to get the right answer to some service problems.

Of all the heating system service problems I have encountered, I would say that the most common one is lack of combustion air. There are two reasons for this. When the original installation is made in a new house, it is very seldom that you see any provision made for adequate combustion air. It may be an oversight because of ignorance or it may be for a reason that is very logical at the time. There are so many openings around loosely fitting basement windows and doors that a separate opening for combustion air is not really required.

Linear crackage openings, that allow combustion air to enter, have a way of disappearing after a family buys a house and moves in. About the second time someone feels the draft from that cold outside air moving across the face or ankles, out will come the caulking gun or there will be a hurried trip to the hardware store for weather stripping material. Before you know it, they are congratulating themselves for having improved the livability of the place by correcting the poor work of the home builder.

You have to give the homeowner credit for being a good handyman and following the advice of the hardware store man in doing such a good job. Unfortunately, no one has given any thought to the fact that the furnace is now starved for the air needed to properly burn the fuel and is relentlessly gulping more fuel, producing more flue blocking soot and more carbon monoxide pollution fumes, in its struggle to satisfy that Simon Legree on the wall known as the thermostat.

The day will soon arrive when the homeowner decides that something has gone wrong with the furnace and calls in a serviceman. Here is where the second reason for trouble enters the picture. Is the serviceman one who **thinks** and asks **why** did this happen? Or, is he one who sees a dirty furnace, cleans it and gets it in operation again, then tells the homeowner everything is fine now, but be sure you call us every year so we can get your furnace cleaned and properly adjusted before the heating season starts.

A serviceman who **thinks** will recognize the cause of the problem and the inherent danger to the home and the family in it. Deaths from asphyxiation due to carbon monoxide and blocked flues are frequently reported in the newspapers. Fires from lack of adequate combustion air are common, although it is not easy to pinpoint this as the cause after the fire. It has been demonstrated that oxygen starvation will cause a fire to leave a furnace and travel until it finds the small opening where air is entering and burn just above this point. Those who have seen this demonstration agree that it is a frightening thing to watch.

The serviceman who recognizes the cause of the problem realizes that a source of combustion air is needed. His first question: how much air? For the absolute minimum requirement, use the 10 to 1 rule. It takes 10 cubic feet of air to burn 1 cubic foot of gas. One cubic foot of gas produces roughly 1,000 Btu. If the furnace has an **input** rating of 120,000 Btu, it is going to require 120 times 10, or 1200 cubic feet of combustion air per hour.

The next question: how big an opening will I need to supply this 1200 cubic feet of air per hour? Follow this 10 to 1 rule. Ten cubic feet of air per hour requires 1 square inch of opening. Your 1200 cubic feet of air will require an opening of 120 square inches. Sounds like a lot, and it is, even when you realize this is only a 10 by 12 inch opening.

Where are you going to get this opening? A simple way is to take a pane of glass out of one of the basement windows and replace it with a louver to keep out rain. You may have to add a screen to the louver and, in some cases, provision for an air filter on the inside. You will have to be careful about the location of this louver because winter air entering here could be cold enough to freeze water or drain pipes close to it. In some cases it is necessary to fasten the window permanently closed and duct the air down to the floor or across to the vicinity of the furnace.

If windows are not available, wall openings may be the answer. I have heard of cases where combustion air was obtained through the roof by means of insulated ducts to the furnace or furnace room. Motorized dampers, for combustion air openings that open only when the burner is operating, are common on large commercial operations and there is no reason for not using them on the smaller jobs if they are needed.

On one service call, the reason for a soot clogged furnace was traced to the fact that the homeowner had plugged the ventilation openings in the crawl space adjacent to the basement. It kept out the field mice, but it also kept out the combustion air. He had a simple solution. He opened the damper in the basement fireplace. On the next service call, it was necessary to inspect the heat exchanger for corrosion or cracks that would admit gases to the room air because the housewife was finding greasy, sooty dust on the floors. The heat exchanger was in good condition. Investigation revealed the fireplace damper was too small to admit enough combustion air. In addition, there was another fireplace in the first floor living room. Flue gases from the furnace chimney were being drawn back down into the house through the fireplace chimneys next to it.

After the fireplace dampers were closed and a louver was installed in a 10 by 16 basement window opening, the floors had only normal household dust on them and they also reported that the furnace filters were not turning black as they formerly did.

It requires awareness of the problem and **thinking** on the part of the serviceman. He will probably have to do some selling to convince the customer of the need for a combustion air opening. The serviceman certainly should be aware of the problem and put forth the effort necessary in this case. Where his customers are concerned, this is literally a matter of life and death.

9

Caring for Condensers

At the beginning of each cooling season, when you go out to start up the air conditioning systems for your customers, do you ever wonder how many frozen and burst condensers you will find? Usually you do not know the condenser is damaged until the water is all through the system. If it is a hermetic compressor, you have a large sales job ahead of you—to sell the customer on this repair. If it is a warranty job or a service contract your outfit shut down last fall, you have a sizable expense that you may have to absorb yourself. You resolve to see that this does not happen again, but, in the rush to shut down in the fall, there is always the chance that it will happen again.

I remember an installation consisting of 26 package units ranging from 5 to 30 tons each, where nine 7½-ton condensers were found to have frozen during the winter shutdown. The installation was still in the warranty period so it was up to the contractor to make the repairs. After replacing the tube-in-tube counterflow condensers, rebuilding the semi-hermetic compressors, and evacuating, dehydrating and recharging the systems, he presented us with a bill for some $2700. He had drained the condensers on the fall shutdown and did not see why he should be held responsible if the condensers trapped water and kept it from draining. The manufacturer's manual gave the procedure for draining and he had followed the instructions. This brought the manufacturer into the picture and the local branch office arranged for the chief engineer from the package unit division to sit in on the discussion.

The contractor had set up one of the condensers on a bench in his shop and had several liquid measures on hand. He demonstrated how so much water was required to fill the condenser and about one-third of it would not drain out by gravity even with the top opening left unplugged. Air pressure was required to remove more water and even then there was sufficient water left to collect and burst a tube if it collected in one spot. The discussion was quite lengthy but it ended with the manufacturer picking up the repair bill.

As supervisor of maintenance, it was my job to see that this did not occur again. Everyone seemed very relieved to hand this problem to me. A year before this happened, I had had three chilled water coils on air handling units freeze up and burst the tubes when some pneumatic low-limit thermostats had failed. These were nondrainable coils and, on the shutdown, we had tried to blow all the water out with air. We learned that we could blow the water out several days in a row and still get a sizable amount every day. Since they were each 50-ton coils, the recommendation that we fill them with glycol antifreeze for the winter proved to be a costly solution.

The retired Navy chief who was the store maintenance head wanted to dilute the remaining water with antifreeze and we finally figured out how to do this. After blowing out the water with air for at least one hour, we hooked up a pump and filled the coils with pure antifreeze. We circulated this for 30 minutes and then drained it back into the barrel and used air to get out as much as possible. This was checked for freezing point and used in the next coil. Watching our hydrometer and adding antifreeze as needed, we were able to protect five of these 50-ton coils to 40 degrees below zero and use less than twenty gallons of antifreeze.

We used the same procedure on the condensers and were able to protect all of them with less than thirty gallons of antifreeze. If these condensers and the water lines are closed up after the job is done, I see no reason why a cheaper alcohol base antifreeze could not be used, unless there is a possibility of corrosion from the alcohol. If you have any problems of this nature, you might try this method. It might save you some money. More to the point, it might make money for you.

RETUBING HEAT EXCHANGERS

When the serviceman finds that some of the tubes of the larger water-cooled systems have frozen and burst, he is faced with the necessity of removing and replacing these tubes. The procedure is not difficult and can be done if you know the proper procedure and have the proper tools. If you have never done it before, don't try it yourself. Call in an experienced serviceman and have him do the job and teach you at the same time.

Most of the larger sized shell-and-tube heat exchangers consist of a cylindrical steel shell with thick steel tube sheets welded to each end. These tube sheets are drilled to accommodate the copper heat transfer tubes running between them and extending slightly beyond the face of the tube sheets. These tubes are expanded into the openings in the tube sheets to form a gas-tight fit. No sealants of any type are used. It is strictly a metal-to-metal seal.

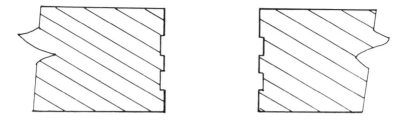

FIG. 9-1 Cross sectional view of tube sheet showing grooves for gripping tubes.

Since copper has a high thermal expansion and contraction rate, it follows that the tubes will try to work back and forth in the tube sheets as the temperature causes the tube to alternately lengthen and shorten. To prevent this action from loosening the tubes in the tube sheet openings, grooves are rolled, or cut, in these openings after they are drilled. They are only a few thousandths of an inch deep, but the gripping action of the grooves effectively anchors the tubes against axial movement. Fig. 9-1 represents a cross sectioned view of a tube sheet opening with two grooves in it. Fig. 9-2 shows how this looks with the copper tube expanded into the opening and into the grooves. You will notice that the expansion of the tube does **not** reach the full depth of the tube sheet. Since the tube is anchored to prevent

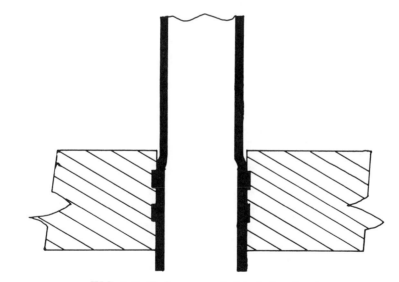

FIG. 9-2 Tube expanded in tube sheet.

movement, expansion forces now cause the tube to bow. This sideways flexing of the tube against the sharp inner edge of the hole could cause a break at this point. Two or three thousandths clearance, for a depth of about ⅛ in., will prevent this

The first step in retubing is to remove the old tubes **without damaging** the openings in the tube sheets. This requires special tube **lifting** tools. You do **not cut** tubes out of a tube sheet, you lift them out with smooth, rounded-edge tools. Fig. 9-3 shows some of my tube-lifting tools with the ends reflected in a mirror to emphasize the smooth, rounded surfaces. All of these tools were made by grinding chisels to the proper shape and then hand filing them for a smooth surface. With the exception of the second tool from the right, none of them have a sharp edge.

To remove the tubes, the rounded end of the tool on the left is placed on the side of the tube, where it extends beyond the face of the tube sheet, and tapped to collapse the tube inwards. This will pull the tube away from the wall of the opening in the tube sheet for a short distance. The point of one of the four lifters can then be dipped in oil

FIG. 9-3 Assorted tube lifting tools.

and placed in this opening between the tube and the hole wall. Holding the straight side of the lifter parallel to the hole wall, hammer the lifter straight in to collapse the tube inward as shown in Fig. 9-4. I file a slight slope on the tops of these tools so they cannot gouge into the hole wall. Usually I start with one of the short lifters and finish up with the long lifter towards the left. The collapsed section must extend beyond the inner wall of the tube sheet. The tool on the right has a flat V-notch in the end and is placed parallel with the face of the tube sheet to drive the end of the tube together where it extends beyond the tube face. The finned condenser tubes in Fig. 9-4 show how these tubes should look after collapsing. You will need to start the tubes out of the tube sheet at one end with a round, flat-end drift; so drive the tube ends together as shown on the left, on this end. The end the tubes are coming out need only be as shown in the second from the left. I use a short-handled, 3-lb. sledge-type hammer for this work.

No matter how long you do this type of work, you will, on occasion, cut through a tube with the lifting tool. You can tell by the sound whether you are collapsing or cutting a tube. A collapsing tube rings with each hammer stroke. A cutting stroke sounds dead. When this

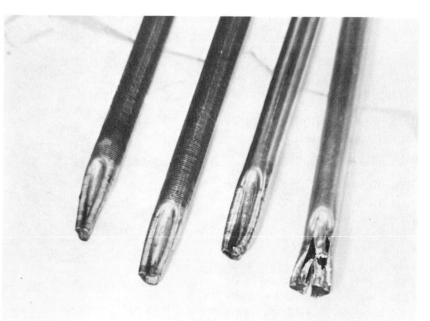

FIG. 9-4 Tubes after collapsing.

happens, take the pick, second from the right, and, with an 8-oz. hammer, pick the edge of the copper tube away from the hole wall and down to form a new pocket for starting the lifting tool again. The tube on the right in Fig. 9-4 has been slit with a hacksaw and spread apart to show where I went through the tube with a lifter and then started over again after forming a new pocket with the pick. The end of the pick is shaped and used, just as a sharp finger nail would be, to pick the tube away from the wall to form a pocket.

Before putting in the new tubes, you can clean the inside of the vessel, if necessary. Clean the face of tube sheet with a wire wheel on a drill. Clean the tube holes. I find that a wire fitting brush can be used to do this. Cut the handle off and chuck the brush in a drill. The tube holes must be clean bright metal to meet with the copper and form a tight seal. Wipe off the new tubes and place all of them in the exchanger. This way, you insure that you have the holes matched at the end of each tube. On long vessels you have to thread these tubes through matching holes in internal tube-support sheets.

Before expanding the tubes into the tube sheet, they must be positioned so that they extend equal distances from the face of the tube sheets. This can be a problem if the tube sheets have eroded faces, shown in Fig. 9-5. This is a four-pass condenser using well water that smelled as if it had some sulphides in it. The red brass, lo-fin tubes were 84 in. long and the

FIG. 9-5 Eroded tube sheet.

tube sheets were 83.5 in. face-to-face. I took an 18-in. tool steel bar, taped ¼-in. key stock to the ends and clamped this to the gasket faces with C-clamps. With the bar at one end of a row of tubes, I

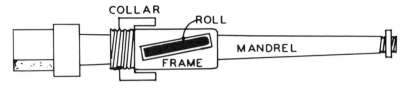

FIG. 9-6 Tube expander nomenclature.

was ready to expand that row of tubes at the other end.

Before going any further, let us see what a tube expander is and how it works. Fig. 9-6 names the parts of a tube expander. The tapered mandrel has a square drive shank on one end and a round or hex-nut keeper on the small end. The frame slides over the mandrel and has slots (3 or 4) for the hardened steel rolls. These are at an angle to the long axis of the frame. At the rear of the frame is a collar that can be adjusted for the proper depth of roll penetration into the tubes and locked in this position. When the expander is inserted into the tube the action is as shown in Fig. 9-7. Clockwise rotation of the mandrel (center) bearing against the rolls (small circles) causes them to rotate in a counterclockwise direction. Since these rolls are bearing against the tube (solid outer ring) the rolls and the frame rotate in a clockwise direction also, although at a slower rate due to the difference in diameter of the mandrel and the rolls.

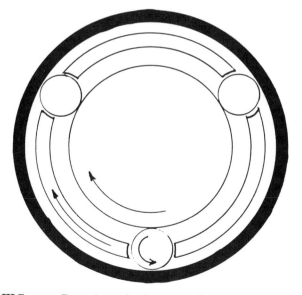

FIG. 9-7 Rotation of tube expander elements.

To see how this action expands the tube, we must refer to Fig. 9-8 showing the relationship between the roll and the mandrel. Lines AC and DE are at right angles to the roll and represent the distance the roll travels in one revolution. Lines AB and DF are at right angles to the mandrel and the same length. Since the roll is not parallel to the mandrel it tries to move backwards along the mandrel as it rotates. The roll is in the frame and the frame is in the tube with the collar against the end of the tube. Unable to move, the roll applies this force (line EF) to the mandrel and gradually draws the mandrel in as it rotates. This increases the diameter of the roll circle and expands the tube.

The roll is also at an angle to the tube and tries to move the distance BC into the tube as it rotates. Since the frame collar prevents

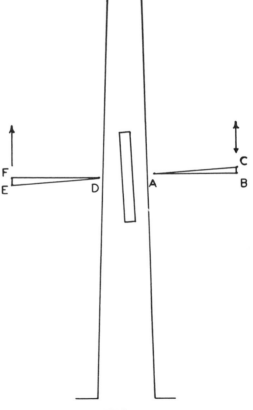

FIG. 9-8

this movement, the force is transferred to a cold working or flowing of the copper in the direction CB by the rolls.

This working of the copper tube is what forces it into the grooves and forms a tight joint between the tube and the tube sheet metal.

If we are lucky, the manufacturer has told us what tube expander to use. If not, we will need telescoping gauges and a micrometer to tell us how much expansion is needed. Fig. 9-9 shows three tube expanders, two telescoping gauges and a micrometer. In the condenser shown in Fig. 9-5, the telescoping gauge inserted in the tube holes *miked* out an average hole diameter of .758 inches. The red brass finned tubes were .750 in. OD at the smooth ends. Wall thickness was .035 in. This meant that we had to expand .008 in. to contact the tubes and then approximately .010 in. more to work the tube into the groves. Trying our expander (center in Fig. 9-9) we found that running the mandrel in to ⅜ in. from the frame gave us a reading inside the expanded tube of .698 in. .758 minus .698 left .060 or a wall thickness of .030 inches in the tubes. Since this was satisfactory, we cut a ½ in. OD copper tube bushing ⅜ in. long and slipped it over the mandrel between the frame and the shank. We were using a ½ in. reversible electric drill with a 7/16 in., 8-point socket to drive the expander. One man runs the drill, the second man inserts and removes the expander from the tubes. **Dip the expander in refrigeration oil** before inserting in each tube.

Fig. 9-10 shows the condenser tube sheet after we had completed the job. In some of the tubes, you can see the rings caused by the grooves in the tube holes. This system was pressure tested at 250 psi

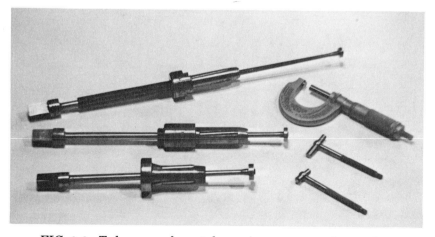

FIG. 9-9   Tube expanders, telescoping gauges, micrometer.

(R-12 with nitrogen) and we had to reroll five tubes. The reflections in the photograph are from an epoxy paint that was brushed on the tube sheet in the hope of preventing further erosion of the steel surface.

BRAZED TUBE CONDENSERS

The serviceman working on water-cooled equipment may run into condensers where the tubes are brazed to the tube sheet instead of being rolled in. The following description of a service call on just such equipment shows how the trouble developed and how the repairs were made.

A 55-ton packaged water chiller developed high head pressures in its third summer of operation. The service contractor circulated acid through the condensers and lowered the head pressures somewhat. He left powdered acid on the job and instructed the help to add a small can of the acid to the tower sump every day. We were called in when two of the four systems in the package were losing R-22 at a steady rate.

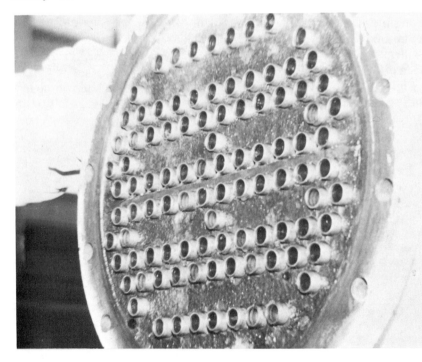

FIG. 9-10 Retubed condenser.

We found refrigerant traces in the condenser water, so we pulled the heads on one end of the four condensers to find out just where the leaks were. These condensers have a ⅜ in. steel tube sheet. The condenser tubes are placed in the tube sheets with a ring of brazing alloy around each tube and then furnace brazed to the tube sheets, Fig. 9-11. This makes an effective gas-tight bond between the tubes and the tube sheets. Since inhibitors do not protect steel from acid, the acid the service company supplied had eaten into the steel and undercut the brazing alloy around some of the tubes.

We valved off the two leaking condensers and removed them from the package. These were taken out and sandblasted to remove the scale and rust from the tube sheets. Incidentally, we found out that a three-second blast removed all scale from the interior of these ⅝ in. OD by 6-foot tubes.

To repair these condensers, we stood them on end and blocked to make the tube sheet level. The surface was flooded with acid flux containing zinc chloride and thoroughly brushed with an acid brush to remove any remaining traces of scale.

FIG. 9-11 Eroded brazed tube condenser.

Using a medium tip oxyacetylene torch with a carburizing flame, we started tinning the ends of the tubes with 430 degree silver bearing solder. Once you have a ring of solder around all the tubes in one segment, you can start flowing solder over the steel. The zinc chloride in the flux will form a black dross on the surface. This is all right, as long as it is fluid. If it starts to dry, add some more flux from the squirt bottle. Apply just enough heat to keep a puddle of molten solder in a small area. Once you have this puddle, move the torch away and rub the steel surface with the end of the solder wire to tin the steel. Do not overheat and do not cover too large an area at a time. Continue working until you have covered the entire area, including the gasket area, with a layer of solder. Now wash this with a fine spray from a water hose and remove all the black dross. You can inspect your work and with flux, torch and solder, you can touch up any parts that don't look right and flow the solder to a smooth uniform surface. The 15-ton condenser heads, shown in Fig. 9-12, required about 1 lb. of solder for each head.

FIG. 9-12 Tube sheet after soldering.

Leak test with nitrogen at a pressure below the relief valve setting or at the nameplate test pressure. Lay a thick soap or detergent liquid and water layer over the work and wait. If you have any leaks, a fine bubble area will show up. It may take 10 minutes for this to occur. Leaks will be around a tube or at the junction of the steel and solder at the perimeter. We soldered all four condensers on both ends and feel that this thick solder layer will prevent any further tube sheet erosion in normal service.

Since the thick solder layer increases the overall length of the condensers, it was necessary to slot the bolt holes in the condenser mounting channels when reinstalling them. Each condenser was purged and an activated alumina drier and a moisture indicator installed in the liquid line. The one wet system required three dryers to show *dry*.

Head pressures are now normal and a constant bleed is on the tower water line. The customer was forced to operate with only one-half capacity for 24 hours until the first condenser was reinstalled. Then he had 75 percent capacity for two days and 100 percent on the fourth day. His saving over the cost of new condensers was considerable.

CLEANING CONDENSERS

Cooling water conditioning and treatment processes reduce scale and corrosion problems, but seldom do away with the need for cleaning the condensers during the off season. Water treatment, constant bleed, and condenser cleaning are processes that go together. It is important that the serviceman understand this and even more important that he get this fact across to the customer. Treatment and constant bleed combine to keep the equipment operating at the lowest possible operating cost during the season. Condenser fouling increases the head pressure and this means more power is needed to operate. Cleaning the condensers means you start the season operating at the lowest possible costs. It usually costs more for the increased power than it does for the treatment.

The need for condenser cleaning should be determined before the end of the season. Find the TD between leaving refrigerant and condenser water temperatures. Compare this with readings taken at the beginning and during the operating season. Equipment should be operating at as near as possible equal load conditions. Condenser fouling will show up as a gradually increasing TD. This procedure is important on tube-in-tube and shell-and-coil condensers. It is not always possible to visually inspect these types for scale.

Shell-and-tube condensers have removable heads that allow visual inspection. You can see the fouling and determine its type and get a better idea of what cleaning process, or processes, is necessary. You can also check the tube sheet and heads for corrosion and erosion problems. Preventive maintenance at these points is often necessary.

Condenser water is usually passed back and forth through the tubes several times. The number of passes is determined by the design of the partitions in the heads. These partitions press on the gasket of the tube sheet and separate the tubes into groups that are covered by each partitioned area in the head. Half the tubes in a partitioned area bring the water into the area; the other half take the water out. The respective position of the partitioned areas in the two heads governs the flow from partitioned area to partitioned area. You can see from this that the two condenser heads must be installed correctly. When you remove these heads, note the markings on them, or mark them, to insure that they are replaced correctly.

Corrosion is always possible in these heads. Erosion is an even bigger problem. The water must make a 180 degree turn in this space and has a very abrasive effect on the metal. Check the thickness of the metal in the partitions. If you find much wear and excessive thinning of the partition metal, repairs are necessary. You may be able to get replacement heads but they are sometimes not available and often expensive. They can be repaired by brazing or arc welding with a cast-iron or stainless electrode. This is not easy to do and my personal preference is for epoxy repairs.

The epoxy repair kits sold by ACH&R wholesalers will bond well with cast-iron heads and holds up very well in these heads. These kits are small and fairly expensive. You may want to buy larger quantities from plastic supply houses. Those specializing in boat and automotive glass fiber materials are usually the best. Start by having the heads sandblasted clean. Make sure that all the rust is gone and keep the heads clean, dry, and free of oil after they are cleaned.

Mix the epoxy according to directions and use small quantities. Epoxy can be built up in layers with good adhesion between layers if you clean each layer with MEK solvent. Glass fiber cloth can be used as a reinforcement to build up thick layers. To get a true flat surface of the face that meets the gasket, lay the head face down on a smooth surface covered with waxed paper while it cures. Final surfaces can be shaped and smoothed with an oiled tool. Table knives and spoons make good smoothing tools. Cover the entire inside surface of the head with an epoxy layer and you will stop all corrosion of these surfaces. Curing the epoxy can be done in an oven for small heads. Larger heads can be warmed with infrared heat lamps. You can shape and trim the cured epoxy with sand cloth and woodcutting tools, if necessary. The smooth surface of the epoxy eases the turning water and seems to be very erosion resistant.

Tube sheets, the face where the tubes are rolled in, are also subject to corrosion and erosion. Some condensers have universal gaskets. These are punched to fit over the tube sheets with an opening for each tube. They fit on the condenser regardless of the partition pattern of the different heads. Tube sheet surfaces under the gasket area that does not contact the head seem to breed a very rapid and deep type of corrosion. Corrosion in crevices around the tubes is usually the worst. You may want to leak test the tubes and reroll them if necessary, if you find very bad crevice corrosion.

Corroded tube sheets can be epoxy repaired if you can get them clean enough. Sandblasting is the best cleaning method. If this is not possible, wire brushing with a power brush, acid cleaning, and a final cleaning with MEK, may get the surface clean enough for epoxy. Check the surface with a magnifying glass to make sure it is clean. Failure to get all the surfaces clean, including the crevices, will make the repair ineffective. Build the epoxy up to the desired level. Before it cures, wax the head surface heavily and bolt the head up to the tube sheet, without a gasket, to get a true surface on the tube sheet repair.

Visual inspection of the tubes will tell you what needs to be done here. Any loose or soft deposits should be brushed out before they have a chance to dry. Use a nylon bristle brush of the proper size. Most supply houses carry them. If not, check with a mill supply house or your Fuller Brush Man. Washing should accompany the brushing. On long tubes, you may want to use $\frac{1}{8}$-inch pipe as a brush holder and connect a hose to the pipe.

There are times when deposits are too thick to push a brush through. This often happens if the tower is not kept clean and a lot of dirt circulates with the water. In this case, brush cleaning before acid cleaning is a must. Cleaning with a rotating brush is necessary. Use a slow-speed drill. I attach a right angle drive adapter to my half-inch drill. Quarter or three-eighths rod gives me the length to reach through the tubes. Slow forward speed and frequent removal, with a jet of water in the tubes between brush insertion, will clean these tubes.

Never, never, use a twist drill to clean condenser tubes. You may have to start out with a small diameter brush and clean again with larger diameter brushes to do the job on some condensers, but at least you will not cut through the tube wall this way. Just compare the cost of a brushing job to a tube replacement and new refrigerant charge: your cost, not the customer's.

ACID CLEANING

When the serviceman starts in to acid clean a condenser he must think and plan ahead. Always establish circulation with clear water before you add acid. See Fig. 9-13.

Look at the piping layout. Does it have any air traps in it? Some condensers, especially centrifugals, drop the water down as it passes through. Acid cleaning creates gases as the scale dissolves. If these gases become trapped in pockets, they will prevent the acid solution from dissolving scale in these pockets. You may have to circulate in reverse flow. You may have to install vents at the high spots to release gases. Look the situation over and think before you start hooking up your equipment.

Once you are hooked up, start circulating the water and prove that you have adequate circulation and no leaks. Plain water leaks are not as damaging as acid water leaks. Note the water level in your container. Keep it fairly low. This leaves room in case trapped gases force the level to raise. **Never add all the acid at once.** Add the acid slowly and make sure it has an inhibitor mixed with it before adding.

Acid solutions eat metal as well as scale. Inhibitors are added to slow down the action of the acid on the metal. They do a very good job on copper and most brass alloys, but are only mildly helpful on iron and steel. Special inhibitors are available for galvanized metals. The fact that iron is so hard to protect with inhibitors makes it imperative that you look the installation over carefully before acid

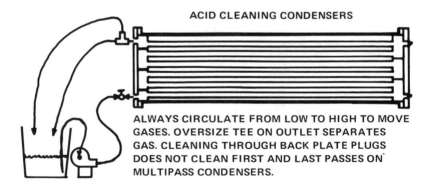

ACID CLEANING CONDENSERS

ALWAYS CIRCULATE FROM LOW TO HIGH TO MOVE GASES. OVERSIZE TEE ON OUTLET SEPARATES GAS. CLEANING THROUGH BACK PLATE PLUGS DOES NOT CLEAN FIRST AND LAST PASSES ON MULTIPASS CONDENSERS.

FIG. 9-13

cleaning. If any of the piping is black iron, you stand an excellent chance of eating a hole in it. The aerated water produced by a cooling tower will contain plenty of dissolved oxygen, and scale formation and oxygen will cause local action cells on the interior surfaces of black iron pipe.

These local action cells will be the first to lose their scale cover with acid cleaning and the thin metal wall left in these areas will be the points that develop leaks when the solution eats through them. Look at the tower construction. Is it asbestos cement board? You can be sure that acid solutions will attack this material. Whenever possible, use a separate pump for acid circulation and confine the cleaning to the condenser and adjacent piping. If scale and corrosion in the piping are reducing flow, be sure the customer understands that you may uncover leaks in cleaning the pipe. There are times when it is cheaper to replace piping than it is to risk damage from acid solutions.

The acid cleaning liquids and dry powders that are available from your supply house have the inhibitors mixed in them. Some are only for iron and copper use. If you want galvanizing protection it will cost more. You can buy muriatic acid and inhibitors from chemical supply houses and mix them yourself. It will pay you to do this when using large quantities. Compare costs before you decide. I prefer to buy ready-mixed acid cleaners from the wholesaler. Storage of these chemicals is a problem and few servicemen have space for this storage. It must be kept away from tools and equipment when it is stored.

Do not make your solution too strong. A pH of 3 will actually dissolve scale faster than pH of 1. This seeming inconsistency is because the strong solutions can produce coatings on the scale that form a barrier and slow down further action. Keep your solution close to pH of 3 by checking at intervals and adding small amounts of acid. By doing this, you will have a better idea of how the descaling process is working. Variations in foaming of the return solution have absolutely no value as an indicator of descaling progress. A piping hookup that properly vents the gases will not return foam to the container. By holding the solution pH where you can detect variations as small as 1 pH and adding acids in small amounts, you will be able to tell when the acid is no longer being neutralized by the scale. When the pH holds steady for an hour, or two hours at the most, it is time to drain the system and make a visual check.

Always neutralize your solution before draining and discarding it. Use soda ash for this and add it while circulating until a pH of 7 is reached. Never leave solution in the equipment overnight. If you find it necessary to continue acid cleaning another day, drain and refill with a solution of clear water and soda ash. Circulate this at least one half hour. An alkaline solution like this will work on phosphate or silica scale overnight and make the acid solutions more effective the next day. Drain and refill with water before starting the new acid solution.

If you must dump acid in the basin and leave it to circulate unattended, be sure you take out plenty of liability insurance first. Damage from acid solution leaks can be expensive. Acid solution in towers and sprayed from nozzles can cause the buildup of large quantities of acid foam. Wind currents can carry this foam off a roof and drop it on motor cars. The car owner's policy does not cover this damage, your policy does. Acid solutions that are drained without first being neutralized can damage sewer lines and pollute waterways. Think about these things.

Tube-in-tube condensers and shell-and-coil condensers cannot be brushed out before acidizing. Strong acid solutions can sometimes cause scale to come off in large chunks that may block the flow. The need to start with solutions that are weak is apparent in these cases.

Venting trapped gases may also be a problem. You may need to provide for reverse flow in some cases. Valves in the lines will help. You may also need valves to keep the solution from draining back and overflowing the container of solution on shutdown. Test for this before you add the acid.

Some condensers wash greasy soot from the air and equipment located near kitchen exhaust vents will deposit oily substances in the tubes. If cleaning action is poor and you find an oily scum on the solution in the container, add a detergent to the solution. I have used liquid *Vel* and powdered *Dash*, also *Spic and Span* powder. Add these slowly and in small amounts to prevent excessive foaming. They do not seem to interfere with the acid cleaning action and they do remove the oily scum that may be coating the tubes.

10

Using Instruments

Service instruments are just more tools for the serviceman's use. Complicated tools, or more intricate tools, or more helpful tools, depending on how you look at them. Useless tools and a waste of money, if the serviceman does not know how to use them correctly, has

FIG. 10-1 Multimeter.

not been trained in their proper use, or does not put them to their proper use. They can even be dangerous tools in some instances.

A serviceman who understands how to use a neon test light in checking circuits can easily solve a problem and arrive at the solution while the man who does not know the proper use of a multimeter, Fig. 10-1, is still struggling away. I have seen this happen more than once. I still remember the time when I set my resistance thermometer knob on *adj* and watched as the needle settled at a point about one degree above 32. A service manager with me pulled out his pocket screwdriver and moved the zero adjustment on the instrument face to bring the needle to 32. Although I knew he owned a similar meter, he had just shown me that he had never read the instructions, and did not understand how the instrument worked.

Zero needle adjustment is used to match the needle with the zero point when the instrument is turned off and not in use. I believe this is true of every instrument I own, regardless of its purpose. The recalibration screw on the dial of my pressure gauges is also a zero adjustment. Always make sure that the small pulsation damping screw

FIG. 10-2   Resistance thermometer.

in the gauge connection is clean and not holding back any pressure in the bourdon tube before zeroing gauges.

My Simpson 389 3L thermometer, Fig. 10-2, has two screws under top caps that are marked *adj* and *cal.* The *adj* screw is the one that must be used to set the needle on 32 when the knob is on *adj.* The *cal* screw is used to calibrate the instrument when the probe is in a media of known temperature and the instrument is set to the proper scale. Fill a vacuum bottle with a mixture of crushed ice and water

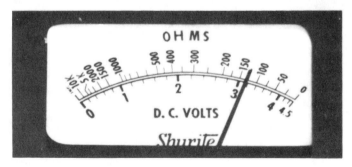

FIG. 10-3   DC volts / ohms scale.

and mix it well. Place the probe in the mixture and set the knob on *low temp.* Calibration is necessary if the needle does not settle on 32. You can check mercury or spirit thermometers by this same method. You can compare the instrument readings to thermometer readings at other temperatures by using water in a vacuum bottle to hold a steady water temperature.

Zero adjustment may be required if the instrument has had some rough handling and *adj* may need changing as the battery voltage changes. Calibration should be checked at least twice a year. It is a good idea to check your glass tube thermometers at the same time. If you know how much error is present in the readings, they are still useable for your purposes.

An ohmmeter is a very useful instrument as has been mentioned before. An ohmmeter is nothing more than a DC voltmeter with a different dial on it. You can buy meters with both scales on the face, Fig. 10-3. Following Ohm's Law, anytime you put resistance in the circuit, you reduce the voltage, Fig. 10-4. The ohms scale is calibrated to show the resistance that corresponds to the new voltage.

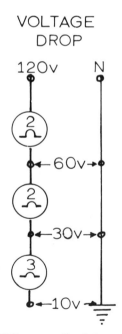

VOLTAGE DROP

FIG. 10-4 Resistance
reduces voltage.

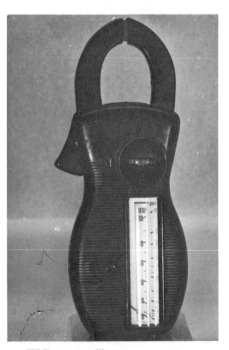

FIG. 10-5 Clip-on ammeter.

A meter of the type shown is under five dollars in a radio supply shop. Add a 0-50 ohm variable resistor and three flashlight batteries and you can make a useful ohmmeter. Three D-cells, in series, gives slightly over 4.5 volts DC. The potentiometer allows you to zero the needle. If you can find a meter with a 0-1.5 volt range it will have an ohms range with a wide scale. This would be useful in checking pure resistance in hermetic motor windings. Special purpose instruments such as these are not too expensive when you make them yourself. The first voltmeter, ammeter, ohmmeter, that I used in service work were made in this way.

The service instruments available to the serviceman today are extremely versatile in relation to the small size. The Amprobe clip-on ammeter, Fig. 10-5, is so easy to use when compared to my ammeters that had to be wired into the line. Scales can be changed to cover a wide range of current, voltage, and ohms. You can see the instruments holding them in almost any position, but I suggest that you check the zero setting of the needle before you use it in that position.

A multimeter such as the Simpson 260 has many ranges of resistance, voltage, (both AC and DC) and amperage that are helpful to a serviceman who is working on a wide variety and size of equipment.

The serviceman working on a limited range of equipment can obtain smaller, less costly instruments that will be adequate for his work.

The Simpson 390, volt, amp, wattmeter shown in Fig. 10-6 is one of the best service tools I have for use on domestic refrigerators, freezers, and window air conditioners. Nameplates and service manuals give the starting and running wattages, or amperages, for the units. Plug the instrument in and then plug the unit in. You will have to make special adaptors for the special types of plugs now used. Voltage is always shown and wattage or amperage is read by just pressing the proper button. Most service manuals give detailed instructions for analyzing

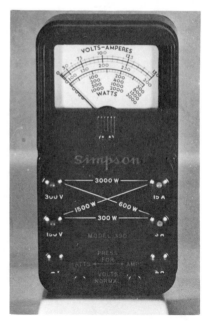

FIG. 10-6 Simpson wattmeter.

unit performance and diagnosing service problems from starting and running wattage. The Frigidaire equipment manual is a good example of this.

The three lead resistance thermometer can be used as a diagnostic tool on domestic equipment with equal results. A General Electric service manual gives procedure and method of analyzing equipment performance with this instrument. The serviceman who has both instruments, **and knows how to use them,** can make very accurate diagnosis of service problems. The three lead resistance thermometer has an almost unlimited range of use in the industry. Temperature is temperature, regardless of whether you are checking superheat on a window air conditioner, or a 6,000-ton centrifugal.

Electronic leak detectors are expensive but there are times when they are the cheapest solution to a problem. The disadvantage to their use is that they are a waste of time and money **unless they are properly used and maintained.** The man who uses this instrument must be thoroughly familiar with the operation of the unit. He must know how to maintain it, how to adjust it for sensitivity, and how to use it.

If you buy one for yourself, read the instructions. Read them again. Try it out on several installations and read the instructions each time to make sure you are using it correctly. If you are buying the instrument for use by the servicemen in your shop, train the men in using it. Make sure they know how to use it properly and keep it in good operating condition.

MEGGERS

One instrument that you seldom find in a serviceman's truck is a megohmmeter, Fig. 10-7. It is more commonly called a *megger.* Its purpose is to test the insulation resistance between motor windings and motor frame. Since it usually measures the resistance in millions of ohms (megohms), it is called a megohmmeter. An ohmmeter will also measure resistance in megohms but where an ohmmeter applies 1.5 to 6 volts direct current and measures the voltage drop on a scale calibrated in ohms, a *megger* applies from 250 to 10,000 volts direct current, depending on the type of instrument. I believe you will agree that a motor that is safe at 6 volts might not be safe when you operate it at 440 volts.

The megger is not a cheap instrument. Unless the serviceman is actually working on monthly inspection work with preventive maintenance in mind, the cost might not be justifiable. Where it is used, and detects even one weakness that could have resulted in a burnout, it will pay for itself right then. The megger shown is also an ohmmeter and AC-DC voltmeter that has proven to be very accurate. Complaints to power companies of high and low voltages, based on this meter, have been checked out and not contested by them.

The serviceman who is using a *megger* in preventive maintenance work **must keep accurate records,** Fig. 10-8. Equipment temperatures and operating conditions should also be recorded. Properly kept records are the arrow pointing to future trouble.

Several years ago I found that I needed a *megger*. The company I worked for had recently had motor burnouts on two 150-hp hermetic centrifugals and the high cost of repairs prompted a memo to us to be on the lookout for this type of damage in our preventive maintenance work. Now *meggers* are not cheap and you don't run down to the supply house and pick one up over the counter in our town. Checking the catalogs, I found several varieties, sizes, and prices, and I decided to investigate before I spent that much money. I obtained AIEE Bulletin #43 from the American Institute of Electrical Engineers. The title of this bulletin is: *Recommended practice for testing insula-*

FIG. 10-7   Megohmmeter or megger.

tion resistance of rotating machinery. It costs one dollar. Their address is: 33 West 39th Street, New York City. I certainly recommend that you read this before you invest in a *megger*.

The following quotation is from paragraph 4.42 of the bulletin: "For purposes of standardization, 60 second application of 500 volts direct current is recommended where short time single readings are to be made on windings and where comparisons with earlier and later data are to be made."

I finally bought a Model 201 *Vibrotest* made by Associated Research Inc. of Chicago. Since it is battery powered, it is easy to use and supplies a constant 500 volt DC potential for test purposes without the danger of burnouts from a high amperage supply

My next question was, what is the minimum test value that is safe for hermetic motors? This turned out to be quite a problem. After much reading and correspondence, I found the two sheets which are shown here, Figs. 10-9 and 10-10. Very little has been written on megohmmeter testing of hermetic compressor motors. My experience has shown that the Westinghouse table is a very useful guide for the serviceman.

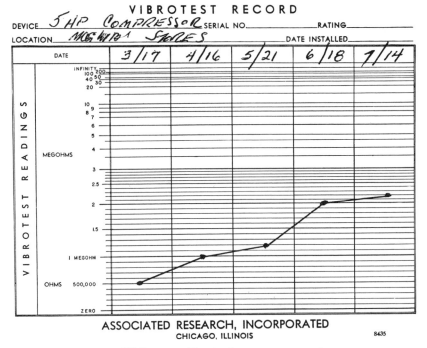

FIG. 10-8 Typical megger record.

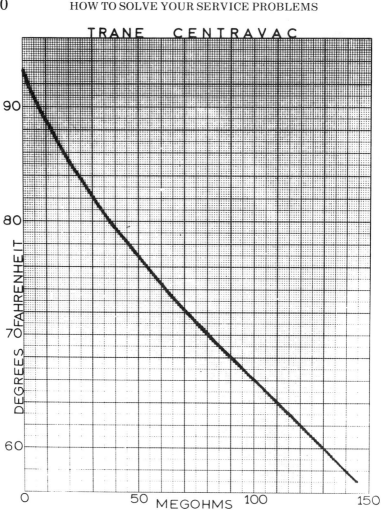

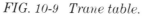

FIG. 10-9   Trane table.

HOW TO READ TRANE CENTRAVAC MEGOHM CHART

1. If test value of insulation resistance lies to the right of the curve, the motor is acceptable.

2. If test value lies to the left, the motor must be operated with caution. The insulation resistance should be checked hourly until it has definitely been established that the insulation resistance is increasing. If it is decreasing, the motor should not be operated.

3. The motor should not even be started if the insulation resistance is less than 5 megohms.

An ohmmeter will show you a grounded winding in a motor if the ground is heavy enough. I can recall one instance where two 135-hp hermetic reciprocating compressors were checked for grounded windings before the initial startup with a 1.5 volt ohmmeter. Both checked okay. On startup, one of them burned out instantly. A 500 volt check would probably have caught this defect and saved the installer a lot of grief.

Another time, I used a *megger* to check the windings of a 100-hp reciprocating job that had been blowing an occasional 400 ampere main fuse for no apparent reason. The *megger* indicated zero resis-

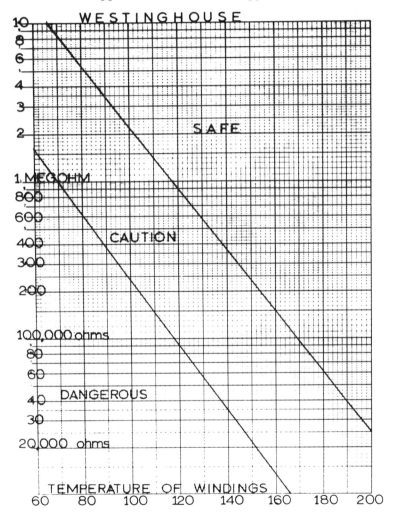

FIG. 10-10 Westinghouse table.

tance and an ohmmeter finally showed 4000 ohms to ground with 1.5 volts. We did not run the machine again until the stator had been rewound. If you have ever had a burnout on a 16-cylinder Westinghouse, Model CLS 3440, you will know how much we saved in repair expense.

In using the *megger*, always try to take the readings after the compressor has been in operation for at least one hour. This will give you a uniform temperature index for comparison readings. If you check a cold machine before startup, using R-22, you can expect to find low readings. Quoting a Westinghouse service letter: "R-22 has an unusual effect on the resistance of the motor insulation and very often can lead you to believe the motor is bad when, in reality, it is in quite good shape." My records on 42 hermetic compressors in Missouri, Iowa and Minnesota using R-22, showed readings from 50 megohms to infinity during the summer operating season, yet the same compressors showed readings as low as .1 megohm during the winter shutdown. Before you condemn one of these compressors, warm the compressor with heat lamps and take another reading.

Remember, if you take a reading without disconnecting the starter leads, you are also testing the wiring and the hermetic terminal block. If you get a low reading, it will be necessary to disconnect the terminal block leads and check again. Another low reading means you must open the compressor and disconnect the winding leads from the terminal block. One low reading on a 200-hp Carrier centrifugal finally pinpointed two defective terminal block insulators as the cause.

I have found that *megger* readings on a hermetic compressor motor will always improve at a steady upward rate after a drier has been installed in the system. Since driers are very good acid neutralizers, this tends to bear out a theory of mine. With the constant expansion and contraction from alternate heating and cooling of the windings, it is possible for minute cracks to develop in winding insulation. Acid in oil or refrigerant films over the windings would cause short circuiting that could lead to burnouts. I make it a practice to install or change the drier core on any hermetic system where I find a questionable megohmmeter reading. It seems to pay off in much less burnout troubles.

BURNOUTS

There is no easy way to check a hermetic motor with electrical test instruments and say definitely that it is a *burnout*. If a lead wire burns apart, you can say definitely that there is an *open* circuit. If insulation burns off and bare wires contact the iron laminates or frame, you can say definitely that the wiring is *grounded*. If coil windings burn and fuse together without grounding, the motor will not have electromagnetic capability around the burned coils. This cannot be checked externally. It will reduce the pure resistance of the burned coil, but there is no practical way to check this in the field.

The serviceman faced with a possible burnout can check for opens and grounds. If he has a *megger*, he can check for loss of motor insulation value. If he still lacks convincing proof, he must then check for the presence of burnout products in the system. A burnout produces carbon and acids. The carbon is a visible product, and can be seen in the refrigerant, in the oil, and on internal surfaces. The acids can be smelled and found with simple chemical tests.

Testing for a burnout only comes after a compressor has refused to start or run and the serviceman has checked out the capacitors, relays, safety controls, supply voltage, and wiring continuity. If no internal burnout products are found, and all external start and run devices are in order, then you can try a startup again. Make sure the proper size fuse or circuit breaker is in use and let the startup pro-

FIG. 10-11 Amprobe Deca-Tran.

ceed until the compressor either starts or the safety opens. If it is no go, repeat the external and internal checks and tests.

The serviceman must be sure that the problem is an internal one before he sends the hermetic back to the manufacturer. If they test and find it was an external fault, the exchange will not be in warranty. Cost of the exchange compressor will be charged to the dealer or serviceman. Make sure the cause is an internal fault or burnout for your own protection.

Don't overlook the accessories that can be used to extend the range and usefulness of your instruments. The Amprobe *Deca-Tran* Fig. 10-11 enables you to use the standard instrument on motors with high amperage ratings. There is also an accessory that can be used to multiply low amperage readings.

The small test lights shown in Fig. 10-12 are neon pilot lights in a panel socket-type holder. Rated at 220 volts, they can be used as test lights for circuits with power on. Wire these lights across the terminals of a solenoid valve and they will light only when the valve is energized. They make good temporary indicators when you are checking out the operation of a step controller or other control in the system.

Know your instruments, their capabilities, their uses, their construction. Check their calibration at least twice a year. If repairs are

FIG. 10-12 Neon test lights.

needed, your wholesaler can send them back to the factory for service and repair. Know the limitations of your instruments too. The 390 wattmeter can be used as a check instrument on a capacitor, but I would be very careful when doing this. If the capacitor is accidently discharged while connected to the instrument, you have a badly damaged instrument. Mine has been back for repairs twice for this reason. I will not check capacitors with it anymore.

Know when **not to use an instrument.** Never use an ohmmeter to check microcircuits. Your ohmmeter voltage may be anywhere from 1.5 to 90 volts. The sensing circuits of some controls, especially heating controls, are designed to operate in millivolt or microvolt ranges. Ohmmeter voltages can and do burn up these circuits if applied across them. Special instruments are required for these controls.

CAPACITORS

A capacitor consists of two large area sheets of foil separated by a thin semi-insulator, called a dielectric. Electrons can pass through these dielectrics and, if you will connect a dry-cell battery to a capacitor, it will absorb electrons on one sheet of foil until it is saturated with all the extra electrons the sheet can hold. You have, in effect, created a storage magnet. You have charged the capacitor. A capacitor differs from a storage battery in that a capacitor will always discharge all of its energy instantly when it is connected to a circuit to do work. This is a useful phenomenon to a photographer who can take two or three minutes to charge a capacitor from dry-cell batteries and then discharge the capacitor in a fraction of a second through a high-voltage flash lamp.

As long as a capacitor is taking a charge, it will not pass voltage. Connect a capacitor to AC current and it will take a charge as long as the electromotive force (EMF) is building to its peak. The instant the peak is reached, EMF starts to flow through the capacitor.

Starting capacitors are usually the electrolytic type. The dielectric between the foil sheets allows fast charging at the expense of some heating. They must not be left connected very long or they will overheat and burn out. Running capacitors usually have oil-type dielectrics that charge more slowly and do not allow the heat to build up in the capacitor. They are also larger than starting capacitors.

Only three things usually go wrong with a capacitor. They can open, they can short, or they can deteriorate. Before checking any capacitor, be sure it is discharged. A capacitor is a storage box of electric energy that condenses its power so that all of it is discharged at once. After making sure that the power supply is disconnected, short across the terminals to discharge the capacitor. Don't use your good screwdriver for this. The potential is strong enough to ruin it. 330 or 440 volts from a capacitor on a 220 line, arcing across, can melt particles of metal or even stop your heart.

An open capacitor will not deflect the needle of an ohmmeter applied across the terminals. A shorted capacitor will instantly deflect the needle to show no resistance. Both good and deteriorated capacitors will start out with no needle deflection and gradually build up resistance as the capacitor takes a charge. Deterioration of either the plate area or of the dielectric between the plates reduces the microfarad (mfd) capacity. Check the capacitor with your ohmmeter **set on the Rx1 dial.** Most small motor capacitors have too much resistance to be checked at higher settings. If the capacitor does not show open or shorted, you can try a replacement capacitor to see if it will cure the trouble. You can check the capacitor with a capacitor analyzer. Capacitor deterioration is hard to find without the special instruments designed for this purpose. Charging the capacitor and then seeing if it will spark when you short across the terminals will not give you any measurement of microfarad capacity. If *mfd* is too low, it will not do the job for you.

Capacitance depends upon area of the plates in the capacitor and also upon the thickness of the dielectric. If you are stuck out on a job with a bunch of capacitors that are not the right size needed, it may be possible for you to substitute a combination of other capacitors in an emergency such as this. Combining capacitors in series reduces total capacitance. You will have to do some paper work on this. The formula for series capacitors is as follows:

Total capacitance equals: Microfarads 1 times microfarads 2
                                   divided by
                                   Microfarads 1 plus microfarads 2

Divide the result of the multiplication by the result of the addition to obtain total capacitance of capacitors in series. Capacitors in parallel: total capacitance equals the sum of the individual capacitances added together. Total capacitance equals mfd 1 plus mfd 2 plus mfd 3. Capacitors in series increase dielectric thickness. Capacitors in parallel increase plate area.

11

Circuits and Controls

An electrical connection is any joint between conductors of electrical energy that is supposed to form a path for current flow. **There should be no resistance in a connection,** The effectiveness of the connection depends upon the area of contact and the closeness of contact. Watch a good electrician make a pigtail splice and you will see him cut back the insulation, scrape the wire to remove oxidation, and then twist the wires together with his *Kleins* to form a tight twist about an inch long. He will then cut the end off so that both wires are equal in length, and wrap the insulation very tightly around the splice. The first layer of insulation is a soft rubber *splicing compound* that conforms to the surface of the splice to keep all air away from the surface. The second layer is a tough vinyl tape that squeezes the compound tightly and forms a tough exterior surface.

The scraping cleans the surface down to bare metal. The twisting brings a large area of the cleaned metals into close contact. The splicing compound is chemically inert and keeps oxygen in the air from oxidizing the metal in the contact area. The buildup of layers of insulation forms a barrier to energy arcing across to another conductor.

Why worry about oxidation? Oxidation on the surface of a metal acts as an insulator. Oxidized metal has considerable resistance. A good many aluminum foil capacitors use only two sheets of metal foil with oxidized surfaces placed together. The oxidized metal layer is all the dielectric needed in these capacitors.

Oxidized metal has resistance and resistance to energy flow changes energy to heat. Heat increases oxidation, melts and burns insulation. The next thing you know, you have reduced current flow and service problems. From my experience, I can say that poor electrical connections are the cause of well over half of all electrical service calls in our industry.

Mechanical connectors are widely used today. They vary in type. One is a brass spool that slides over the twisted wires, is locked on with a setscrew, and has a bakelite cover that screws onto the spool. Another has a copper spool that slides over the twisted wires and is held in place by three indentations from a special tool. An insulator sleeve covers the spool. Yet another type has a spiral, wire-wound insert in a plastic insulator that threads its way onto the wires to hold them together. Slip-on connectors that allow wires to be connected to matching male/female components on devices are very common. The connectors are crimped to the wires with special crimping pliers.

All of these connectors are good and will do the job they are designed to do **if they are properly sized and applied.** A connector designed for two #14 wires will not connect enough surface when used on two #12's. A connector designed for two #12's will not hold two #14's tightly enough to prevent oxidation between the mating surfaces. A poor crimp, not enough crimps, or oxidized wire crimped in a terminal, will overheat. A wire that is held in place by a terminal post screw that is not sufficiently tightened will eventually oxidize and overheat. I realize that this is only a little thing to make such a big fuss about, but I have learned the hard way that this *little thing* can grow to be a big problem if not corrected.

When I made the first monthly inspection on a large installation in warranty, I made it a point to shut off each unit and check all the electrical connections in the switches and magnetic starters. There were some connections loose enough to cause problems later, but the big difference I found was in the mechanical splice connectors. Even after less than a month's operation, some of them already were showing signs of heat damage. From the location of the connections, it would seem to indicate that at least two electricians were on the job and one of them did not know how to properly install the type of connectors being used. The wires were not inserted into the connec-

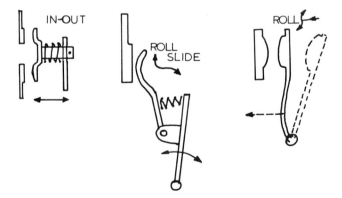

FIG. 11-1 Three types of switch contacts.

tors so that the ends matched and, as a result, the area of contact was very small. All of the poor connections showed visible heat damage within eight months and were repaired without damaging equipment. None showed up after that time. The serviceman who **thinks** and looks for heat damage around terminals and splices will prevent a lot of trouble later on. Experience will teach you what to look for in the way of heat damage.

SWITCH CONTACTS

To control the flow of electrical energy, we must have a means of opening and closing the conductor paths. We do this by installing switches in the circuits, Fig. 11-1. The points of contact in the switches are subjected to the energy forces as they make and break. Since potential exists between the contact points, electric energy can arc across the gap between the points whenever the gap becomes small enough for it to do so. This arcing will damage the contacting surfaces. Pitting can reduce the contacting area so that resistance builds up to reduce the amount of current conducted when the contacts are closed. This results in heating of the contacts and further damage. Sometimes the points will weld together and fail to break contact.

Switches are designed to make and break contact quickly to minimize arcing. The simple toggle switch quickly opens or closes the contacts with spring pressure regardless of how slowly you operate the toggle arm, Fig. 11-2. Once you pass the center of directional force, the movement of the switch contacts is completely controlled by the spring. Early line-voltage control switches used a small permanent

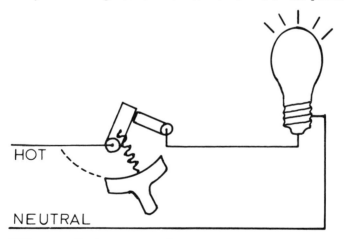

FIG. 11-2 Spring action of single pole toggle switch.

magnet that snapped the contacts together when the iron keeper came close to it and resisted opening of the contacts until opening pressure was great enough to snap the contacts apart.

The glass-enclosed mercury switch is also a snap-action switch, Fig. 11-3. The mercury maintains contact as it is moved towards the other end of the tube by the actuating arm. When the shifting weight of the mercury is enough to overbalance the center of gravity of the tube, the mercury quickly rolls to the low point and moves the tube with it. Enclosing the switch points in an inert atmosphere within the tube prevents oxidation of the surfaces from the arc.

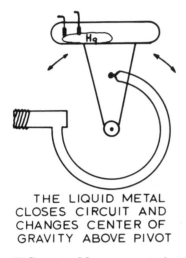

THE LIQUID METAL CLOSES CIRCUIT AND CHANGES CENTER OF GRAVITY ABOVE PIVOT

FIG. 11-3 Mercury switch.

Reducing the potential across the contact points is one reason for low voltage control systems. Low voltage in a control switch will cause little damage to the contact points and this same low voltage can be used to power the magnetic coil that operates the higher voltage switch. This is known as a relay coil. The control system of a residential heating and cooling system is a good example of this. You can even control a thousand-ton system from a 24-volt control circuit, if you will relay more than one time. Amperage across contact points determines the amount of point damage. You can use 24 volts to operate a relay magnetic switch that closes the contact points to energize a 220-volt coil that closes the contact points to allow 2300 volts to energize the motor. The relatively slow movement of the low voltage control is converted to very fast acting control movement in the relay switches. Each relay is able to carry, through its contacts, the greater amount of current needed in the next control system.

You can reduce arcing and heating of contact points as they make and break, but you cannot eliminate it entirely. Most contact points are of silver-plated copper. Silver is a very good conductor and the copper is a very good absorber of heat. The contour of the mating surfaces is designed so that the movable surface either slides or rolls on the fixed surface after contact. This separates the areas of initial contact and allows the heat in this area to dissipate without harming the surfaces. You may melt the silver plating in this area with initial contact, but the film of silver is self-healing as it cools and resumes its original shape.

The thin layer of silver can become dirty and tarnished (oxidized) with time and use. You never clean silver-plated points, **you polish them.** Place a business card or strip of good bond paper between the points, and move it back and forth, while holding the points together if necessary, to polish them. The cotton fibers in the paper are abrasive enough to polish, but do not scratch the surface. Any scratches will slow down electron flow across the surface and localize heat damage. The layer of silver is so thin that even fine crocus cloth can remove it entirely with only a few strokes. The cheaper papers and brown wrapping paper are what is known as sulfite paper. There is enough sulphur residue in this paper to tarnish the silver instead of removing the tarnish. Never use them for this reason.

All but the smallest and cheapest magnetic contactors have replaceable contact points. Electrical wholesalers stock these in replacement kits that include all the accessory items such as springs and flexible leads. When you start having troubles with magnetic contactors due to bad point surfaces, it is time to sell the customer on installing one of these replacements. Do the job right and use all of the parts in the kit.

Failure to replace all contacts, springs, pivot pins, and bushings when overhauling a contactor, could result in the contactor failing to drop out when de-energized. I recall such an occurrence in Omaha on a multiple-compressor water chiller. All safety controls functioned but one motor kept on running. The chiller froze and burst the tubes. Refrigerant was lost and the system filled with water. Three compressors were repairable, but the fourth had to be replaced, as even the outer casting was damaged. All because a magnetic contactor failed to drop out.

CIRCUITS

A circuit can be defined as the pathway of electrical energy from the source to the machine or device and back to the source. Let's start at the generating plant.

There are two leads coming from each phase from the generator. A total of six leads then are fed to the step-up transformers. The step-up transformers could have the secondary windings connected in the Y configuration shown which would result in three hot lines leaving the step-up station for transmission to the locality where power is to be used. A step-down substation would reduce the voltage for distribution to the users where voltage would be further reduced to that needed on the users' circuits.

Generating plants have many different ways of connecting transformers for power distribution. They want to balance the loads as closely as possible on each phase and they want to get the best power factor they can. You will find many different types of distribution and user transformer connections used by the same power company as they try to tailor their power supply to the needs of an area and of the individual user. The drawings show some of the methods of connecting transformers that I have run into.

You will notice that I distinguish between a neutral line and a ground. They are not always the same. It is possible to transmit current through a line that is grounded. The three-wire, three-phase with B-phase grounded is a good example, Fig. 11-4. The fusible switches for this power supply have only two fuses and two disconnects. B-phase is connected to a terminal bar and is not disconnected at any time. You still get three-phase power from this circuit. A voltmeter will show 220 volts from two of the phases to ground. It will show 220 volts between all three phases. It will not show voltage from B-phase to ground because this phase is already grounded. You cannot get 110 volts without installing additional transformers.

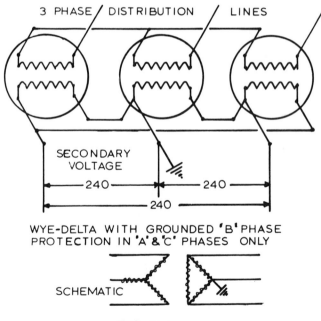

3 PHASE DISTRIBUTION LINES

SECONDARY
VOLTAGE

240   240

240

WYE-DELTA WITH GROUNDED 'B' PHASE
PROTECTION IN 'A' & 'C' PHASES ONLY

SCHEMATIC

FIG. 11-4

You can get 120-volt power from the ungrounded three-phase circuit shown in Fig. 11-5. You will find these types of transformer connections used on older installations where the primary need for electrical power was for three-phase.

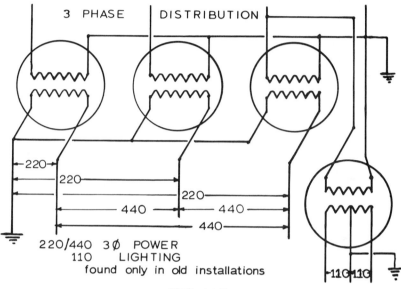

3 PHASE DISTRIBUTION

220/440 3Ø POWER
110 LIGHTING
found only in old installations

FIG. 11-5

The open delta connection that requires only two transformers for three-phase power is also common on older installations, Fig. 11-6. It, too, has a grounded B-phase. This is not a neutral line. This is a grounded circuit.

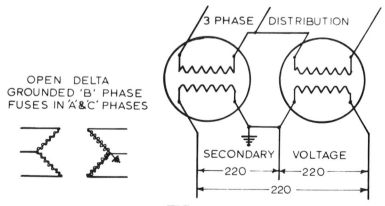

3 PHASE DISTRIBUTION

OPEN DELTA
GROUNDED 'B' PHASE
FUSES IN 'A'&'C' PHASES

SECONDARY VOLTAGE

FIG. 11-6

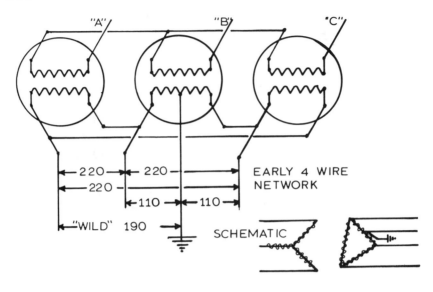

FIG. 11-7

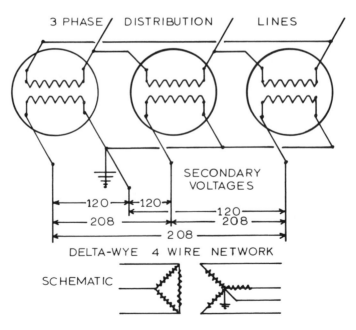

FIG. 11-8

To reduce the number of transformers needed, power companies went to the four-wire network system, Fig. 11-7. It is possible to get two voltages from a transformer by tapping the secondary windings at the midpoint. This is the neutral line we are familiar with. A neutral line is actually a grounded subcircuit line. The delta connection shown was one of the first types of four-wire network systems I had to work with. It was very easy for the power companies to use compared to the standard three-phase connections they were using. The delta connection did not do a good job of balancing the load between phases because it took all its single-phase from one transformer. Single-phase was available from either B or C-phase to neutral. A-phase to neutral was always referred to as the *wild* phase. Its voltage was too low where single-phase, 220 volts was required and caused an awful lot of trouble if someone tried to use it for 110-volt lighting requirements.

The Y-connected, four-wire network system, Fig. 11-8, is the most common one in use today as far as I know. The power companies can do a better job of load balancing and power factor correction with this type of customer useage transformers. The customer's three-phase equipment, of course, must be capable of operating at this voltage.

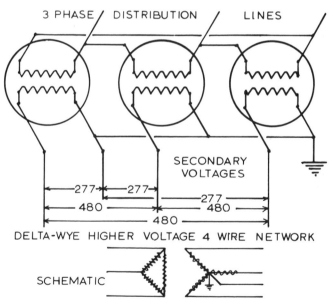

DELTA-WYE HIGHER VOLTAGE 4 WIRE NETWORK

FIG. 11-9

The 277/480 volt network shown in Fig. 11-9 is used in today's commercial buildings. The 277 volts is supplied to the fluorescent lighting and reduces the size and heat output of the lighting fixture transformers (ballasts).

The 480-volt power reduces the wire size necessary for supplying machinery loads. Equipment connected to the 480-volt supply *must* be designed for 480 volts, not 440 volts.

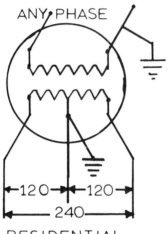

ANY PHASE

←120→←120→

— 240 —

RESIDENTIAL
TRANSFORMER

FIG. 11-10

Air-cooled transformers, located within the building, take 480 volts and reduce it to 120/240 volts for incandescent lighting and office equipment. Wiring is the same as a residential transformer.

Residential lighting and single-phase power is from a single transformer as shown in Fig. 11-10. Air conditioning, heating, electric ranges, and clothes driers use the 240-volt, single-phase power. Lighting loads are divided equally between the two 120-volt lines.

The two 120/240 volt lines are protected at the main entrance. Additional fuses or circuit breakers protect branch lighting and power circuits.

The serviceman should know and think about what happens when a neutral line opens, as shown in Fig. 11-11. With no midpoint circuit, the voltage supplied in either 110-volt hot line can run wild. Turn on a motor in one circuit and the voltage will drop low enough to burn out the motor. At the same time, this voltage drop will cause the voltage in the other circuit to vary from 110 to as high as 190. Light bulbs will flare up and burn out and, in some cases, explode. A voltmeter connected to these lines will show these fluctuations. The 110-volt lines will continue to carry power with a broken neutral because it is the custom to ground the neutral line either at the meter or at the disconnect switch. I have run into this situation only three times. In each case, it was necessary for me to disconnect the house service grounding line before I could convince the power company that the trouble was a broken neutral line at the transformer.

The earth is a ground. It does conduct current but it can never be considered as a conductor of a circuit. And a neutral line is a safe line only as long as it is connected to a ground. The minute you disconnect a neutral line from a ground, the neutral line becomes a hot line, Fig. 11-12. Splices in neutral lines should be made with the same degree of care as any hot-line splice. Mechanical connectors need to be just as tight. Splices need to be insulated to prevent oxidation and resistance buildup in these splices just as much as hot-line splices do. The same amount of current travels through the neutral as does through the hot line. **Think about it. It might save your life.**

SAFETY DEVICES

The conductors in a circuit are sized to carry the amperage needed to do the necessary amount of work. The electrical energy work devices connected to the circuit are designed to use only a certain amount of current. Any accidental groundings or shorting out of resistances in the circuit will increase the current flow. To prevent damage to the conductors or devices, or destruction from uncontrolled electrical energy, safety devices are installed to interrupt the current flow.

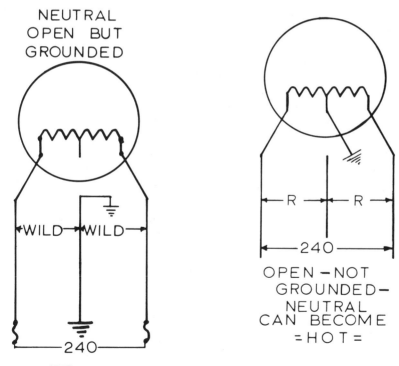

FIG. 11-11   *FIG. 11-12*

The simplest safety device is the one-time fuse: a metal strip, in a container that has a weak link. This weak link will start to heat when excess current passes through it. When the metal gets hot enough, it melts and opens the circuit. The link opens as it melts and creates an arc across the open area that increases as the metal melts back in the electrodes. A short circuit tries to force so much current through the link that it literally explodes apart.

One-time fuses are good protectors on resistance-type circuits, but they do not give as good protection on inductive-load circuits. **Think** about it. An induction motor requires, on the average, three and one half times the running amperage to start it. You can damage a motor if it **runs** and consumes 25 percent more current than it is designed for. The fuse must be sized to carry the much higher starting current. **Therefore, a one-time fuse will not protect a motor that is running.**

To overcome this problem, dual-element fuses are used. A fuse link that can carry the starting current is connected in series with a spring-loaded solder joint that will melt slowly if the current passing through it exceeds the running amperage. A short circuit will open the fuse link instantly. Excess amperage will open the solder joint slowly. The **time** element of a solder joint allows us to have the dual protection necessary in this case.

One thing must be emphasized in the use and operation of fuses. That is, **heat** from electrical energy is used to trip these safety devices. The fuses may be plug types that screw into a socket or they may be cartridge types that insert into clips or knife-type holders. The fuse holders must make good contact with enough area to pass the rated amperage without heating. Any resistance in the fuse contacts will cause the fuse container to soak up this heat and open below the rated amperage. A warm fuse is a source of potential trouble. A heat-damaged fuse clip may lose its spring tension and this means a poor contact. Look for heat damage. Replace damaged fuse holders.

CIRCUIT BREAKERS

Fusible alloys can be compounded to known melting points and electrical resistance characteristics. These alloys can be applied in conjunction with springs and mechanical leverage and operating forces to produce other types of safety devices. Place the alloy between a shaft and a movable latch and you will prevent the latch from tripping as long as the alloy is a solid, but you will trip the latch as soon as the alloy melts. This opens the circuit and stops the current flow by allowing a spring-loaded switch to open. After the alloy cools enough to harden, the latch can be reset to close the switch.

This is the operation of the circuit breaker that is being used more and more to replace the one-time fuse. Nothing needs to be replaced after the breaker trips and the chances of changing the amperage rating by installing a larger fuse is eliminated. You cannot put a penny behind a circuit breaker either.

OVERLOAD RELAYS

An alloy with a known melting point can also be placed between a shaft and hub and heated by an external resistance heat element, Fig. 11-13. In this case there is no electrical connection between the two circuits. The current is passed through the heater coil on its way to the motor. The spring-loaded switch, that is opened when the alloy melts from external heat, is in the motor starter control circuit. We use the heat energy of excess amperage to relay the signal to the control circuit to open. Any device that operates in this fashion is known as an overload relay. The overload heaters come with different amperage ratings to produce the necessary melting temperature. Heaters must be selected to match the connected load plus the degree of overamperage that can be allowed.

It used to be that you could safely allow 115 or 125 percent overload on motors. Not any more. With the price competition and space reduction of today, especially in hermetic motors, allowable overloads may be down near 5 percent. Never install a larger overload

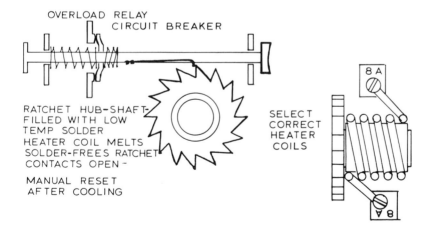

OVERLOAD RELAY
CIRCUIT BREAKER

RATCHET HUB–SHAFT
FILLED WITH LOW
TEMP SOLDER
HEATER COIL MELTS
SOLDER-FREES RATCHET
CONTACTS OPEN –

MANUAL RESET
AFTER COOLING

SELECT
CORRECT
HEATER
COILS

8 A

8

FIG. 11-13

heater than the factory specifications call for. This voids the warranty. Since overload heaters are heat elements, good terminal connections of these heaters is an absolute necessity.

WARP SWITCHES

Metals expand when they are heated and different metals have different rates of expansion. Bond two different metals together in a bimetallic strip and the different rates of expansion will cause the strip to warp as the metals expand.

The warp switch takes electrical energy that has been converted to heat and then converts this heat to movement, Fig. 11-14. The movement can be slow or a snap action, depending on the type and design of the bimetallic element. Tie this movement to one contact of a switch and we can make or break the switch. There are many applications of this type in our industry. Enclose the warp switch in a capsule and bury it in the stator windings of a motor. Excess stator heat will open the switch and break the control circuit of the motor starter. This is an internal protecto-relay.

Form the bimetallic metal into a disc, that is dished slightly like the bottom of an oil can, and place one of the switch contacts in the center of this disc. When the disc is heated, there is a snap action as the expansion force finally inverts the dish shape and vice versa. There is an audible click as this type of switch makes and breaks. The trade name for it is *Klixon*. Attach this switch to the outside of a hermetic compressor and you can operate it from the heat of the

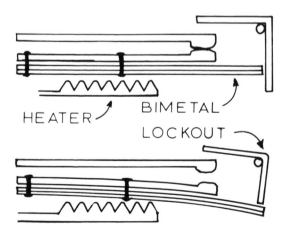

HEATER   BIMETAL

LOCKOUT

FIG. 11-14 Warp switch.

motor. Pass the motor current through a heater positioned near the disc and you can get closer amperage flow control action, Fig. 11-15.

The snap action of this disc makes it possible to run motor current through the switch so that you have a direct-acting safety switch. It can also be used as a relay switch to open and close the control circuit. Contact points can be smaller and less subject to arcing damage when used as a relay. The disc-type warp switch can also be used as a fan switch and a high-limit switch on heating units. Temperature is temperature to a warp switch whether it comes from electricity, gas, coal, or oil. The fact that these switches are nonadjustable is a disadvantage to the serviceman. To the manufacturer who does not want factory control settings altered, it is a distinct advantage.

I have noticed some confusion among servicemen over the terms line-voltage and low-voltage control systems. In our industry, the term line-voltage control usually means that the controls operate at the line voltage and are capable of carrying the **load amperage** through the control contacts. **There is no separate control circuit.** This can be a single switch for a light bulb or a thermostat for a motor. You can add switches to a light to control it from more than one location, as shown in the three-pole switch application and the four-pole switch application, Fig. 11-16. You can add controls to the motor starting thermostat as shown in the compressor application with pressure switches and overload switch added, Fig. 11-17. All of these controls are sized to carry the connected load amperage through their contacts.

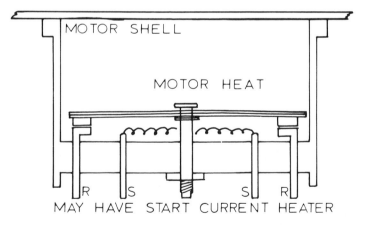

FIG. 11-15 Bimetallic disc motor protector.

MULTIPLE SWITCH CIRCUITS
INDEPENDENT SWITCH ACTION

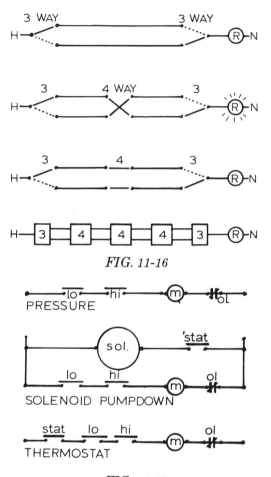

FIG. 11-16

FIG. 11-17

When you get into larger systems with more amperage, you run into much more expensive controls. A control circuit that relays the control commands to magnetic starters does not require contacts capable of carrying the full amperage load, Fig. 11-18. The control circuit may be line voltage or it may be low voltage. The only thing that matters is that it is a separate control circuit. You can do many more things with a separate control circuit than you can with direct-control switches.

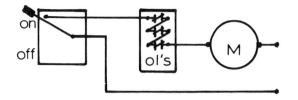

FIG. 11-18 *Manual control of magnetic connector.*

The three-wire, push-button control system in Fig. 11-19 is one in common use for manual control of motors with more safety than an ordinary manual starter gives. A momentary power failure with a manual starter results in the motor immediately restarting. In some applications, this could result in damage. The three-wire, push-button control drops the starter out and it must be manually restarted. This type of control was a motor saver back in the days when we had

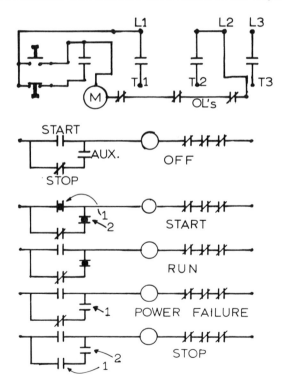

FIG. 11-19 *Three-wire push-button control.*

to manually open the bypass valves on the compressors each time they were started.

The operation is as follows: pressing the momentary contact start button energizes the coil and closes the contacts. The auxiliary contact closes and maintains the coil circuit in series with the normally closed stop button. Either a momentary power failure or a momentary push on the stop button opens the control circuit and drops out the magnetic starter.

Study the one-time pumpout (OTPO) control diagram, Fig. 11-20, and you will see that this is the same control circuit. The thermostat must be calling for the compressor to start. After the thermostat (STAT) opens, the compressor is kept in operation by the one-time pumpout low-pressure switch until this low pressure is reached and the OTPO switch opens. The compressor can cycle on the low pressure (LP) switch as long as the thermostat is calling (closed). The compressor cannot restart until both the OTPO and the STAT are closed. The high-pressure switch (HP) is of the manual reset type.

When trouble occurs in the control circuit or power circuit, it calls for a continuity check. Electrical current will not flow unless it has a continuous path available. A continuity check is a check to find where the circuit is open. Sounds very easy, doesn't it?

Let's take a look at what should be a simple service call on a residential air conditioner that has blown a fuse. This should be a simple trouble for the serviceman to diagnose and repair, but it is surprising how many times it turns out otherwise. The home has a 3-wire, 220-volt entrance drop, 110-volt circuits for lighting, and 220 for water heating, stove, and air conditioner. One fuse is blown between terminals 1-2 in the fusible disconnect.

The serviceman may use an incandescent test lamp in a pigtail socket to locate an electrical problem. If it has a 110-volt lamp in it,

ONE TIME PUMP OUT CONTROL

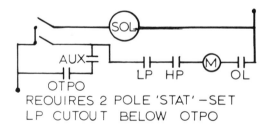

FIG. 11-20

he will be careful not to put it across any terminals that he knows are 220-volt. If he tests 1-5 and 3-5 with the switch closed, Fig. 11-21, the lamp will light. He has proved to himself that he has power in each 110-volt circuit. He then tests from 2-5 and 4-5 and the lamp lights each time. He may think that he has proved there are no blown fuses, but in testing between 2-5 he has actually drawn power for the lamp from terminal 4 through the primary winding of the 24-volt control circuit transformer to terminal 2. The lamp may light or it may not. It may burn bright or it may burn dim. It all depends on the size of the lamp and the amount of resistance in the primary winding of the control transformer. If he puts the lamp across terminals 3-4 it will not light because current will not flow through resistance if it has a path of no resistance around it. If he places the lamp across terminals 1-2 it may or may not light, again depending on the resistance of the primary winding of the control transformer.

If the serviceman uses a neon light he may put one lead of the tester on terminals 1, 2, 3, 4 in turn leaving the other lead open. He will get a dim light from each terminal. If he repeats this test with the other lead touching his finger, he will get a bright light each time. Neon test lights operate from potential instead of current flow and the body acts as a capacitor to give a bright light if potential is avail-

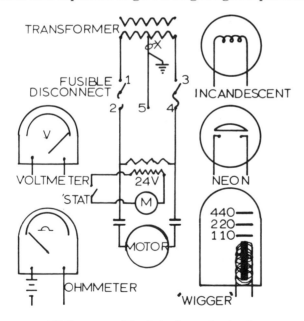

FIG. 11-21 Electrical service tools.

able at the other terminal. If the leads are placed across terminals 3-4, it will not light because only one potential is available. Placed across terminals 1-2, the neon tester will light because of the transformer primary winding. If you know about this transformer and are sure that it has a 220-volt primary winding, then you are sure you have a blown fuse.

But what if the control transformer has a 110-volt primary and is hooked across 2-5 or 4-5? Or, there is no control transformer and the line from 2 dead ends at the open starter contacts? In this case, you will not get a light with either type of lamp across 1-2. You could assume that the fuses are okay and you would be wrong.

The solenoid-type voltage indicator such as the one known as a *Wigger* could be used to check out this problem. Across terminals 1-3 it will have enough power available to bring the indicator up to the 220 mark. Testing between terminal 1-5, 3-5, 4-5 will bring the plunger up to the 110 mark. The voltage between 2-5 will be below 110 because of the resistance of the primary winding in series with the tester leads. If you are sharp enough to spot this reduced voltage and are **thinking** about the fact that you have a transformer winding in the circuit, you will know immediately that fuse 1-2 is the open one. You know there is an open fuse because you got no voltage indication between 2-4. If there is no control transformer and the line from 2 is open at the starter, you will get a no voltage indication from 2-5 and know at once that fuse 1-2 is open. All these ifs mean that you have to **think** while you are using your test instruments.

If you will open the disconnect switch and pull the fuses out of the bottom clips, leaving them in the top clips, then close the switch and test from the bottom of the fuses to 5, you will find the blown fuse at once, because you have eliminated the possibility of any feedback. There is about one chance in a million of this test failing. I'll tell you about that one chance in a moment.

The serviceman's combination clip-on ammeter, voltmeter, ohmmeter, is one of the best tools you can use in solving electrical problems. Testing across terminals 1-3 you will get 220 volts. No voltage across 2-4 will tell you that one or both of the fuses is open. Testing across 3-4 will show no voltage. Testing across 1-2 will show reduced voltage, which means that you have power potential on both sides of the fuse and it must be open or it would not indicate.

If both fuses were blown you would get no voltage across 1-2, 2-5, 3-4, 4-5, or 2-4. You can always remove the fuses, put the battery on the ohmmeter, zero it and check the fuses. A blown fuse should always read infinity on the ohms scale. Well, almost always.

I was called out at 2 a.m. to a plant that processed farm products. A 50-hp. ammonia compressor with a synchronous electric motor driving it had dropped out and plant operators had been unable to get it back on the line. They had called the super, who lived nearby, and he could not get it started. It would single phase every time they tried. Fuses checked okay, so they assumed they had trouble with the electric service and called the power company. The power company servicemen could find nothing wrong with the service and had called in the local district manager to back them up. Things were at a standoff when they called me.

After being briefed by everyone present, I checked the compressor manifold valves, to make sure it was starting unloaded, and closed the switch. It single phased, so I stopped at once. There was no mistaking the noise that motor made. A test lamp showed power on all three phases, phase to phase and phase to ground. I checked to make sure I had no feedback and then got out my clip-on ammeter, voltmeter, ohmmeter.

Power coming into the fuses checked out at 210 volts, phase to phase reading A-B, A-C, B-C. This was good for this 208-volt, 4-wire network system. Voltage leaving the fuses was about 190 for A-B and A-C, and 210 for B-C. Something had to be wrong with A-phase fuse. These were 400-ampere renewable link cartridge fuses. I took the A-phase fuse out and as carefully as possible removed the caps and pulled out the linkage. It had blown, but when it did, a fine thread of wire remained across the gap. This passed enough current to confuse our test instruments, but not enough to have any effect on the motor. If I had not been watching my voltmeter carefully and reading the actual voltage, I would have missed this too.

You have to be thinking all the time when you are using test instruments. I recently took a reading with an ohmmeter across the terminals of a motor starting relay. It read infinity. The homeowner commented, "An open winding, that has to be the trouble." I said, "Not necessarily. The meter is set on Rx1 and we could have an open winding or we could have more than 6 ohms resistance." On Rx100 setting the needle trembled at the infinity mark. On the Rx1000 setting we found the coil was okay and had 2000 ohms resistance in the windings. The trouble turned out to be badly burned contact points in the switch that were not letting starting capacitor current get to the starting winding. We had to replace the relay after all since the contact points were not the replaceable type.

Regardless of what instruments you use in service work, you have to **think** to get the proper results from them.

WELDED CONTACTS

Not all contacts are supposed to be closed when the serviceman is making a continuity check. You have to be thinking when you check across a set of contacts. It is open; is it supposed to be open? It is closed; is it supposed to be closed? There are times when a pair of contact points weld themselves together. This can lead to some confusion. Suppose you have a solenoid keeper that is connected through an insulator strip to operate a dual set of contacts. If one contact has molten metal between the points at the moment they open, they may weld together as they open and lock the contacts together. The other set of contact points may open and be held open because of the angle at which the weld locked into place while cooling.

The arc across a pair of opening contacts may also cause dust and oxide films to solidify into a coating of slag that acts as an insulator when the points close again. You cannot see these things during the usual inspection because there is no room between the points for a visual inspection. This is where the ohmmeter check can be of great value. Don't just use needle deflection as an open or closed circuit check. Zero your ohmmeter and set it on the lowest scale. A closed set of contacts should have no resistance at all. Finding any resistance calls for a closer check to find out why you have the resistance.

The experienced serviceman learns that the same service troubles seem to come in cycles. You will usually find damaged contacts and burned out magnet coils on many calls after a locally heavy lightning and thunder storm. Such storms will affect the voltage of the transmission and distribution lines in the area. A period of over and under voltage during a storm will bring on a rash of troubles in electrical equipment. The serviceman who recognizes this will know what to expect and where to start looking for troubles after a storm.

Look at your surrounding area when you find you have control circuit troubles. Is the equipment located in the rear of a grocery or drug store and not too clean? See any cockroaches? A cockroach wing is a very good insulator when it is caught between two contact points. A bug mashed between the mating areas of a starter magnet will cause extra current demand and can burn out a holding coil. A coal bin near the equipment will cause a fine deposit of coal dust on the contact points that will extend the arc across opening points and buildup of a glossy coating of slag on the points, insulating slag.

Look at your controls. Are they open controls? Do they have enclosures with tight covers? Are the covers in place? Are there any knockouts in the box that were knocked out and not used? **Think** when you look at these controls and at the surrounding area.

CONTROL CIRCUITS

There are many types and designs of control circuits. Unless the serviceman is restricted to a very limited make and model of equipment in his service, he will find it impossible to remember the exact design of the control circuits. The serviceman should always refer to the control circuit drawing when checking out a circuit.

A friend once brought in a multipole switch that controlled the operation of a processing machine in his plant. He had been having a lot of trouble with the machine and was firmly convinced that the trouble was in the switch. Fortunately, he had brought the operating manual with him and it contained a good wiring diagram. We spent more than two hours checking out the switch in each position and comparing the circuits with the diagram. One contact, on one position only, showed resistance once in a while.

We took the switch apart a layer at a time and made a sketch of each layer, so that we could reassemble it correctly. In the third layer, we found a silver-plated contact button that had separated from the short brass strip. It could not travel far, but evidently there were times when it did not make good contact. We were able to solder the button back in place and reassemble the switch. We were unable to get any resistance in further testing and it worked perfectly when reinstalled on the machine. We would never in the world have been able to check out the switch without the wiring diagram showing the switch operation in each position.

I was sent out to help a fifth-year apprentice who had replaced a capacitor and relay on a 3-ton air-cooled condensing unit and could not get it to run. He explained what he had done and said that he could not find anything else wrong with the controls. I asked him if he was sure he had all the wires back exactly as they were originally and he said, "Yes." He did not have a wiring diagram available but the position of the wires agreed with his memory of how a capacitor-start motor should operate. They were studying that particular type of motor in night classes at that time.

I went to the truck and got out my manual on that particular model condensing unit. Checking the circuit, wire by wire, we found just one wire on a wrong terminal. Correcting this was all that it took to put the machine in operation. No one can be expected to memorize all the various control circuits that are possible. There is no substitute for the correct wiring diagram!

LOCKOUT RELAYS

A relay, as we use the word in our industry, is an electromagnetic switch. It is used to relay current in control circuits that operate electromagnetic motor starters. A lockout relay in a control circuit can be a puzzle to a serviceman who is not familiar with them. The relay usually operates two sets of contacts, Fig. 11-22. One set is normally open and one set is normally closed. Checking for continuity with a test light and the unit power on, you will be able to get a light from either side of the lockout relay (LOR) coil to ground, yet the relay will not be energized.

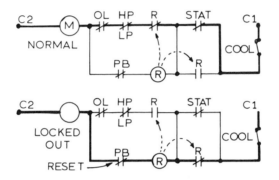

FIG. 11-22 Lockout relay circuit.

This is normal for a lockout relay. To understand why, place a test light across the terminals of a closed switch. The test light will not light because you have only one source of power to both sides of the light. Open the switch. You now have two sources of power and the test light will light. The test light is now in series with the circuit. Another way of saying this would be: current will not flow through resistance if a path of no resistance, bypassing the resistance, is available.

The holding coil of a lockout relay is wired so that it parallels that part of the control circuit containing the safety switches. The normally closed contacts of the LOR are included in the safety switch circuit. The normally open contacts of the LOR are wired so as to bypass the thermostat when closed. As long as all the safety switches are closed, current will bypass the holding coil of the LOR. Current flow will be as shown by the heavy line in the first drawing. When a safety switch opens, current flow will be as shown in the second drawing. The relay locks out unit operation until you manually interrupt current flow to the LOR.

The lockout relay converts every unit with automatic reset safety controls to a manual reset unit. It prevents short cycling that could damage capacitor-start compressor motors. When the unit locks out, it must be manually restarted. This warns the operator to watch it when he restarts it. A second lockout means: find out what the trouble is. Resetting the LOR can be done by means of a normally closed push-button switch in the LOR circuit or by opening the control circuit on/off switch.

Two wiring diagrams are shown here that use lockout relays. The first is of a Westinghouse model RU-52 package unit with push-button reset of the LOR, Fig. 11-23. I have added to the description so that you may more readily follow the elementary diagram as you trace it out in the panel diagram. The second is of a Westinghouse model SU-51 unit with resetting of the LOR done by opening the cooling switch, Fig. 11-24. Description is the same as that shown in my service bulletin. You can see from this diagram that a knowledge of the operation and purpose of a lockout relay is necessary to under-

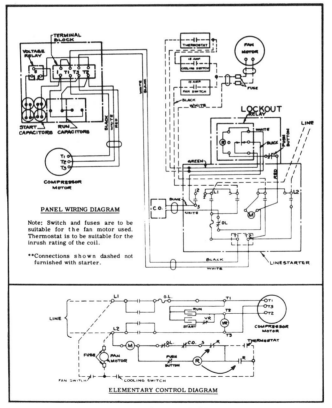

FIG. 11-23 Westinghouse RU-52 schematic.

stand the factory wiring diagram. If you do not understand a diagram, ask the factory man to explain the diagram for you. That is
what he is paid to do.

ADD-ON COOLING CONTROLS

It is very important to study the wiring diagrams when a central
air conditioning unit is added to an existing heating system. The
thermostat for the system is usually replaced with a new one that
has switches for changing from heating to cooling, or keeping both
systems off. A manual or automatic fan switch may also be present.
Trouble comes if the installer fails to remove a red, size 20 wire,
jumper about an inch long on the thermostat subbase, Fig. 11-25.

These thermostats are designed to operate either of the systems
from **one source** of low voltage as they are factory assembled. Most
warm-air furnaces are equipped with a 24-volt control system. Most
of the add-on central air conditioning systems have a 24-volt control
system. This means that each unit has its own control transformer.
If two control transformers are present, the subbase jumper **must be
removed.** Each system will then operate on its own control circuit.

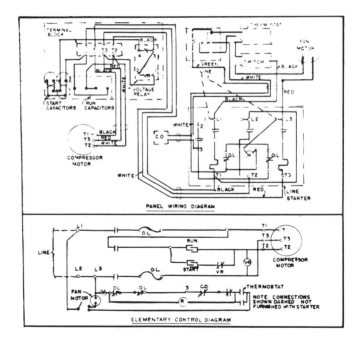

FIG. 11-24   Westinghouse SU-51 schematic.

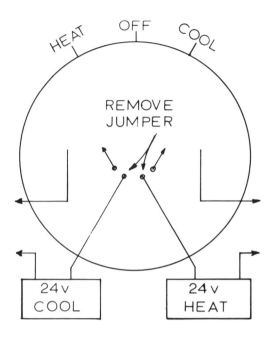

FIG. 11-25 Thermostat adaptation.

Two independently wired transformers feeding into a common circuit from their secondaries will cause a lot of trouble. EMF and current flow must be in phase and it takes special regulating equipment to do this.

The jumper is to be removed only if two control transformers are present. Some add-on systems are matched to the heating system and use the power from the heating system control circuit. The jumper must be used in these cases. I have been on service calls where the trouble was failure to remove the jumper. I have also had calls where the trouble was removal of the jumper without checking the entire control system. When the customer switched over to cooling, nothing happened. Some serviceman or installer **did not think** when he removed that jumper.

CONTROL CIRCUITS

Control circuits can be very simple or they can get quite complicated. The serviceman can trace the control circuit he is checking out if he has a diagram available. The usual panel diagram is not as easy to follow as a line or ladder-type diagram. The Westinghouse elementary control diagram could also be called a line diagram, ladder diagram, or schematic diagram. The schematic diagram is the easiest one to follow if you are trying to understand how the circuit works. The panel diagram is necessary if you are trying to trace the actual wires in the control boxes and panels.

The simplest of all control circuits in refrigeration is the low pressure (LP) temperature control. Since the pressure of the refrigerant varies directly as the temperature, the pressure will decrease as the temperature in the refrigerated space decreases. A low-pressure switch in the suction line can be set to cut off at the pressure corresponding to the desired temperature in the space. The differential can then be set to close the switch when the temperature rises to a given point. In some LP controls the cut-in pressure must be set first and the differential adjusted to open at the desired cutout pressure.

This is low-pressure control. It is simple, direct, and fairly accurate. You can even set the differential wide enough to cut in above freezing and have a defrost cycle each time. With one machine, one coil, one room, pressure control is easy. With more than one lowside on the compressor and different temperatures to contend with, additional controls are required. The usual method was to set the pressure switch to control the coldest space and then install suction pressure regulators in the lines from the warmer rooms to prevent them from getting too cold. If pressure cannot get below a certain point, temperature will not get below that corresponding to the pressure.

Pressure control of temperature works fine on refrigeration as long as the loads are in a fairly close balance. It does not lend itself to air conditioning because we are working at higher pressures and with larger spaces. A residential air conditioning system can be satisfactorily controlled by installing a thermostat in series with the pressure and overload controls, Fig. 11-26. Thermostat control works equally well on refrigerated spaces. Thermostat control does have one major drawback, refrigerant migration on the off cycle. System pressures tend to equalize rapidly and, if the compressor is off long enough to cool down, refrigerant can collect in it.

Thermostat control is preferred on our smaller systems because of the pressure equalization advantage. This makes for easier starting of the hermetic compressors. Crankcase heaters are installed to prevent refrigerant migration to the compressors. Most crankcase heaters are wired in so they are in constant operation. If the power to the unit is disconnected for any length of time, refrigerant can collect in the compressor. Some of the larger units have a separate circuit for the heaters that disconnect them when the motor starter is closed.

SINGLE CONTROL CIRCUIT
MULTIPLE CONTROL

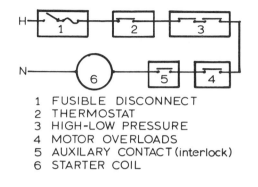

1 FUSIBLE DISCONNECT
2 THERMOSTAT
3 HIGH-LOW PRESSURE
4 MOTOR OVERLOADS
5 AUXILARY CONTACT (interlock)
6 STARTER COIL

FIG. 11-26

Where multiple lowsides are connected to a compressor, pump-down control is often used. Each lowside has its own thermostat and solenoid. Compressor motor control is from the low pressure switch. If wide variations in load are expected, the compressor should have some sort of capacity control. Any leakage through the solenoids, or back through the compressor, will cause the compressor to run long enough to pump down again.

The cycling of a compressor using pumpdown control is objectionable when you get into larger systems. This led to the development of the one-time pumpout circuit (OTPO). The OTPO switch is simply a low-pressure switch added to the control system and wired in series with an auxiliary contact in the motor starter. This circuit bypasses the control thermostat and keeps the motor starter in until the OTPO switch opens. Once opened, the motor starter cannot be restarted until the thermostat again makes the circuit. It is an automatic three-wire, push-button control. The low-pressure switch in the compressor control must be there and should be set to cut out at a lower setting than the OTPO switch. The low-pressure switch is the only low-pressure control that is effective when the thermostat is closed. See Fig. 11-20.

Liquid line solenoids on a OTPO controlled system must either be operated from a double-pole thermostat as shown, or the thermostat must operate a relay in parallel with the solenoid that takes the place of the thermostat switch in the OTPO circuit. You can see that an OTPO circuit would never close the solenoid if it were connected to it.

This control circuit does have a drawback. If the solenoid opens and the compressor circuit stops the compressor for any reason, the open solenoid can flood the system with refrigerant. There is a way to get around this and we will see it a little later on.

OIL PRESSURE SAFETY SWITCH

The oil pressure safety switch (OPSS) is one of the most useful devices for preventing trouble to compressors that I know of. There are times when I think it is the most misunderstood and miswired device in a control system. Basically, it consists of two normally closed switches. One of these switches is always wired in series in the control circuit. It is a warp switch operated by an electric heater that gets its current from the control circuit. It locks open and must be manually reset with a push button. The other normally closed switch is operated by two bellows, one sensing suction pressure and the other oil pressure. Oil pressure must be higher than the suction pressure to open the switch. This switch does not lock out.

The OPSS can only be used on the larger compressors that have a forced feed lubrication system. The setting of the **closing** pressure of the bellows-controlled switch must never be below that recommended by the compressor manufacturer. Always check the manual for their recommendation. The sixteen-cylinder Westinghouse model CLS 3440, for instance, can be operated safely with as little as ten pounds oil pressure. The Carrier model 5H120 twelve-cylinder compressor should not operate with an oil pressure below 35 pounds. It takes this much oil pressure to load all of the cylinders. Normal oil pressure is about 50 psi.

The warp switch that is wired into the control circuit is a time delay device. This allows the compressor to start and build up oil pressure before the warp switch opens. The warp switch heater is energized at the same time the compressor starts. The energizing contacts are closed as long as the compressor is operating. The differential switch is in the heater circuit and, when oil pressure opens it, the warp switch heater stops heating. Loss of oil pressure anytime the compressor is operating will reenergize the heater and open the control circuit and lock it out. The OPSS are available in both line and low voltages and with time delay intervals from ten seconds to two minutes. Automatic reset is also available.

Wiring an OPSS in a control circuit is easily done when you know and understand how it functions. The first drawing, Fig. 11-27, shows how this can be done with a control circuit that is line voltage taken from the motor starter. The circuit also has OTPO. Thermostat and solenoid circuit is separate with switching relayed to the main control circuit. This circuit could result in flooding with liquid from an open solenoid.

The second drawing uses OTPO control of **two** compressors driven by **one** double-shaft motor cooling a water chiller, Fig. 11-28. Control voltage is lower than line voltage. Off cycle liquid flooding is prevented by wiring the solenoid so that it can open only when the compressor motor is operating. The OTPO must have a close differential and fairly high setting with this type of solenoid circuit. Pressure buildup in a new or tight system may not be fast enough to match the ther-

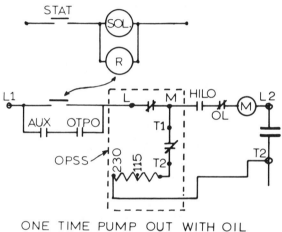

ONE TIME PUMP OUT WITH OIL
PRESSURE SAFETY SWITCH

OPSS HEATER <u>MUST</u> CONNECT TO
<u>LOAD</u> SIDE OF STARTER

FIG. 11-27

NOMENCLATURE

M	Magnetic starter holding coil
S	Solenoid valve electromagnet coil
R	Relay holding coil
OL's	Overload relay switches
HiLo	High and low-pressure cutout switches
L	Line voltage terminals
T	Load voltage terminals
C	Control voltage terminals
AUX	Auxiliary switches in a magnetic contactor or motor starter
STAT	Thermostat controlled switch

mostat cycle. You can see that it is no problem to use two OPSS controls in a single circuit. Since each compressor has its own oil pump, and also shutoff valves, all of the controls shown in this circuit would be necessary from a safety standpoint.

ONE MOTOR—TWO COMPRESSORS—TWO OPSS
OTPO—FAIL SAFE SOLENOID — LOW
CONTROL VOLTAGE

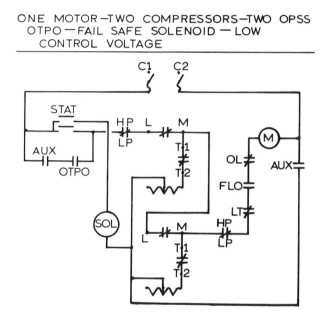

FIG. 11-28

SOLENOID COILS

The centering effect of a magnetic field can be used to operate a valve. The valve disc or needle is connected to a soft iron shaft that can be magnetized when placed in a strong magnetic field, Fig. 11-29. The iron shaft becomes a temporary bar magnet from induced EMF and centers itself in the magnetic field of the coil. Surround the bar magnet with a metal tube in which to travel, made of a good conductor such as brass, and you can keep the electric coil out of the pressure section of the system.

Most solenoid valves make use of the weight of the iron valve shaft to close the valve when the coil is de-energized. This means that the coil must always be installed so that the shaft travels up when energized. Solenoid valves that state in the instruction sheet they can be installed in other than a vertical position have a spring to close the valve when coil is de-energized.

PILOT SOLENOIDS

Small solenoid valves can be used to operate valves of a larger size that depend for their operation on differences in internal pressure. One example is a solenoid valve in the external equalizer line of a thermal expansion valve. You will recall that the external equalizer line is necessary to insure that the pressure on the lower side of a

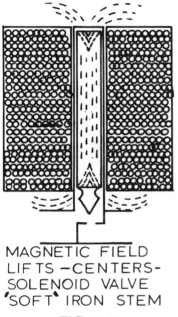

valve diaphragm is the same as the low pressure at the coil outlet. Pressure at the coil inlet is higher because of pressure restriction in the coil. If you close off the equalizer line with a solenoid, coil inlet pressure builds up below the diaphragm and forces the valve to close. The small solenoid acts as a pilot to make a shutoff valve out of the expansion valve. Superheat setting must be low, of course.

MAGNETIC FIELD
LIFTS —CENTERS-
SOLENOID VALVE
'SOFT' IRON STEM

FIG. 11-29

The pilot-operated, diaphragm-type gas valve is another example of differential pressure-operated valves, Fig. 11-30. The diaphragm has two holes in it: a small one that allows entering gas pressure to come into the space above the valve and a larger opening through the pilot solenoid that allows pressure to move out to the low pressure side of the valve. The diaphragm is usually equipped with an assistance spring.

With the solenoid valve closed, pressure gradually builds up above the diaphragm and forces the valve closed. A pressure of a few ounces **per square inch,** is multiplied by the large area of the dia-

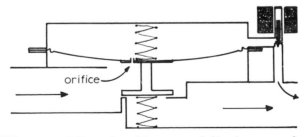

orifice

FIG. 11-30 Pilot solenoid-operated diaphragm valve.

phragm. Opening the solenoid allows pressure above the diaphragm to bleed off faster than it can enter and the valve opens. **Time** allows small differences in pressure to do more work than it would seem possible. Contrast this to the electrical capacitor.

SOLENOID SWITCHES

The same principle that operates a solenoid valve can be used to operate an electric switch, Fig. 11-31. A switch is nothing more than a valve controlling the flow of electricity. Since the switch does not need to be separated from the magnet as a liquid solenoid does, the movable part of the electromagnet can be made so as to come into direct contact with the stationary part. This eliminates the 60 Hz hum that is present in liquid solenoid valves. You will notice that a

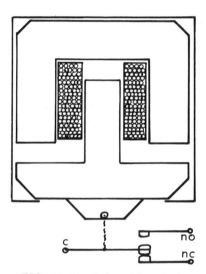

FIG. 11-31 Solenoid switch.

solenoid valve coil is usually much larger than a magnetic starter coil. Since the bar magnet is not touching the primary magnet or a stator, the magnetic field must be larger and stronger to keep it suspended.

Solenoid switches and magnetic starter electromagnets have a divided laminated iron core with the movable part acting as the *keeper* of the magnet. An inrush of current to the coil winding builds a strong magnetic field in and around the core that is concentrated in the area between the magnetic poles. The keeper is located in this field and is drawn into contact instantly by this field.

With the keeper in place, the magnetic field is concentrated in the iron and a smaller current flow is all that is necessary to keep it there. Impedance builds up in the closed circle, or circles, of the magnet to reduce current flow. If anything prevents the keeper from making a good contact with the magnet poles, you will get noise and excess amperage in the coil. The noise is the 60 Hz current cycling from 0 to full EMF. Excess amperage will cause overheating of the coil and can result in burnout. You can remedy a noisy starter coil by trueing the pole faces with a file or sandcloth placed on a flat surface.

BUCK/BOOST TRANSFORMERS

One of the major problems of the serviceman today is that of obtaining the proper supply voltage to insure correct operation of hermetic motor compressor units. The trend to higher supply voltages to buildings such as the 480/270 has not been followed by corresponding motors wound for these voltages. In some cases the serviceman is faced with a supply voltage that is as much as 20 percent off of the nameplate voltage designation. With hermetic motor protection units sized as close as 5 percent in some cases, this results in a lot of service calls. These protective devices are usually sealed and should not be changed, as to do this would cancel the warranty.

The serviceman faced with this problem will not receive much help from engineering, the suppliers, or the power company. The problem originated either because the specifications were not fully read or the equipment was all that was available. In any case, it is easier to pass the buck to the serviceman than it is to come up with an answer. I have had a power company engineer tell me that over voltage would not cause excess current draw in the motor. This is not always true. Both over and under voltage supplied to a motor cause increased current draw (amperage), motor heating and excessive motor noise, unless the motor is built with enough copper and iron in it to operate over a wide voltage range. Over voltage also shortens capacitor life and reduces the power factor of the motor.

Buck/boost transformers are the solution in most cases of over or under voltage. What follows is a description of how they work and how to select and wire in the proper size transformer or transformers.

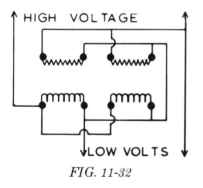

FIG. 11-32

A buck/boost transformer is nominally a low voltage signal transformer. They are dual wound with 120/240 volt primaries and either 12/24 or 16/32 volt secondaries. They are wired into the circuit so that all of the load current flows through the secondary winding before going to the load.

The action of these transformers, when installed for buck boost purposes, is actually that of an auto transformer and is described in most electrical textbooks under that term. Let's take a look at Fig. 11-32 and see how they work.

Boost . . . low voltage is wired in parallel to both the secondary and primary windings. The voltage of the load current flowing through the secondary windings is increased by the *in phase* induction of voltage to the secondary windings from the primary windings. The amount of voltage increase depends on the ratio of turns in the primary/secondary windings and the arrangement of the windings in either series or parallel.

BUCK/BOOST TRANSFORMER SELECTION TABLES

Dual wound with 120/240 primary ratings

	Buck/Boost desired	Secondary Rating	VA Multiplier	Diagram No.
A	*1 phase*	*100/165 volt range*		
	10	12/24	.1	1
	20	12/24	.2	2
	13.3	16/32	.133	1
	26.6	16/32	.266	2
B	*1 phase*	*200/295 volt range*		
	5	12/24	.05	3
	10	12/24	.10	4
	6.6	16/32	.066	3
	13.3	16/32	.133	4
C	*3 phase*	*Grounded B-phase (open delta) 200/295 volt range Two Required*		
	5	12/24	.03	5
	10	12/24	.06	6
	6.6	16/32	.04	5
	13.3	16/32	.08	6
D	*3 phase*	*4 wire network (Wye) 180/280 volt range Three Required*		
	10	12/24	.033	7
	20	12/24	.07	8
	13.3	16/32	.045	7
	26.6	16/32	.09	8
E	*3 phase*	*4 wire network (Wye) 350/500 volt range Three Required*		
	5	12/24	.017	9
	10	12/24	.033	10
	6.6	16/32	.023	9
	13.3	16/32	.045	10

Buck . . . High voltage is wired in series, first to the secondary windings and then to the primary windings. Load current, flowing through the secondary windings first, results in the induction current from the primary windings being slightly *out of phase*. This is the *bucking* action that produces the reduction in voltage.

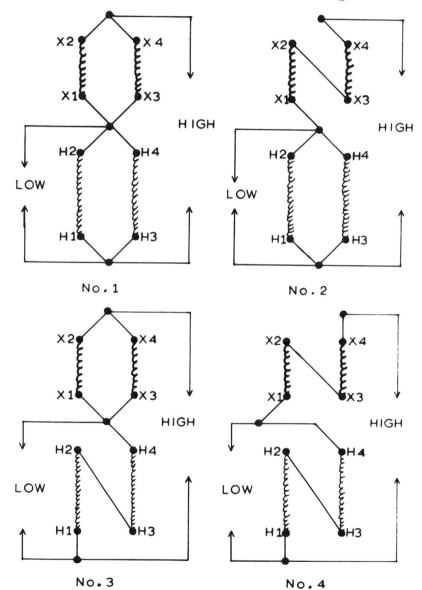

A careful study of Fig. 11-32 will show you that the load current always goes through the secondary windings but never through the primary windings. The **same diagram** is used for both buck and boost. Connect the load to low voltage for buck and to high voltage for boost.

In order to select the proper buck/boost transformer we must first determine the percent of voltage change desired. To do this, subtract the low voltage from the high voltage and divide the result by the low

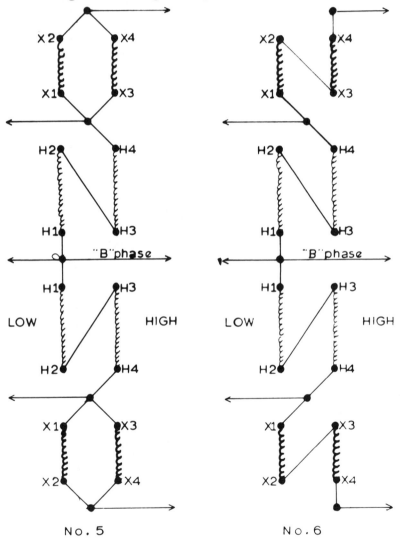

No. 5   No. 6

voltage. Let's take this case as an example: actual supply voltage was 242/248/248 on a three-phase supply with a grounded B-phase. The unit nameplate called for 220 volts, 18.4 amps, 3-phase. 242 minus 220 is 22. 22 divided by 220 is .10 or 10 percent. Use this procedure for both buck and boost. In this particular case a 10 percent buck was required.

The volt-amp capacity of the buck boost transformer depends on the amount of load current and this is not the same on three-phase as on single-phase. For this reason it is necessary to use the selection tables to determine the size of the transformers required. Referring to section C of the tables, a grounded B-phase supply requires two transformers. For a 10 percent change we need a 12/24 volt secondary. The VA multiplier is .06, and we use wiring diagram No. 6. Re-

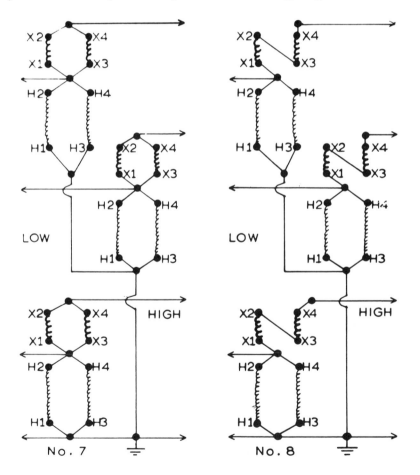

LOW   HIGH   No. 7

LOW   HIGH   No. 8

ferring back to the nameplate, 220 volts times 18.4 amps gives us a VA load of 4048. .06 of 4048 is 243 VA. The closest size in a buck boost transformer is 250 VA.

I installed two 120/240, 12/24, 250 VA transformers for the unit wired according to diagram No. 6. Where the unit had formerly tripped the 20.5 overloads several times a day, it has not tripped out since. Voltage readings are now 222/224/224 and my *Amprobe* indicates load current is slightly under 18 amps.

The transformers cost about $16.00 each. This is extremely low compared to the cost of a motor compressor on this unit. The saving in service calls alone will pay for the buck boost installation in less than a month.

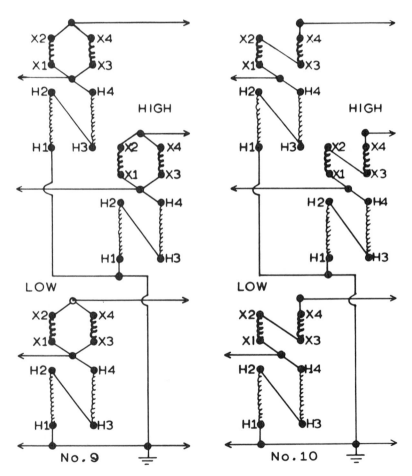

No. 9   No. 10

12

Evacuation and Driers

When you are leak testing a system, you add refrigerant to the system before you add nitrogen to build up pressure. Never make the mistake of trying to add refrigerant after the system has been pressurized with nitrogen. The chance of forcing nitrogen into the refrigerant cylinder is too great. Blow the nitrogen pressure down first. After all, the more dry nitrogen you blow through the system, the drier it gets. You are helping yourself in the next step of the installation which is evacuation.

Evacuation is aimed at doing two things: removing air from the system and removing moisture from the system. Air is a noncondensable and takes up space in the condenser that is needed for refrigerant condensing. We have been over this earlier. Moisture in a system can collect in the expansion valve, freeze, and block valve operation.

The combination of moisture and air in a system can also cause another form of trouble. Fluorinated hydrocarbon refrigerants supply the basic chemical components of some very strong acids. All that is needed for the action to produce these acids is water, oxygen, and a catalyst. The best catalyst is hot steel. The head chamber of a compressor is an ideal catalytic chamber. Once the acids are formed, they wind up in the oil and start attacking the bearings. This is one of the reasons for evacuating the system of air and moisture.

No vacuum pump will remove all of the air and moisture from a system. You can get the amounts left in the system so low that they are harmless. Every stroke of the pump pulls a quantity of air from the system equal to the pump's displacement. The remaining air expands. When you finally reach the point where the pump will no longer pull air from the system, the remaining air is so light in density that, when it is compressed in the operating system, it takes up very little space.

Moisture is removed from the system by reducing the pressure and causing it to vaporize. As long as the metal parts in the system, that the moisture collects on, are warm enough to supply the heat of vaporization, the moisture will vaporize. Vaporization pressures and temperatures are in the table of properties of water. If you pull a vacuum below 4.6 mm Hg you will freeze the moisture. Unless the ice is in contact with warm metal, it may stay frozen for a long time. The fact that you have removed most of the air surrounding the ice means that heat cannot be transmitted through the air. A vacuum is a very good insulator.

If the system contains any oil that could form a film over liquid water, this oil film can prevent the water from evaporating. This is something you must think about if you are working on a system that has been in operation and has circulated oil through it. Liquid water will also be present in a system because of **adsorption.** Adsorption is the ability of a surface to collect very fine droplets of water on it and hold these droplets to the surface by molecular attraction. The rougher the surface, the more moisture can be adsorbed on it.

These are the reasons that I have always preferred what is called the triple evacuation method. After you have pulled a vacuum down close to the freezing point of water, break the vacuum by introducing dry gas to the system to bring the pressure to about 2 pounds gauge. This dry gas can be nitrogen on a large installation or the operating refrigerant on a small installation. Repeat this procedure. Repeat this procedure a third time, but this time break the vacuum with the operating refrigerant. Evacuate once more and then you are ready to charge the system with refrigerant.

Breaking the vacuum with dry gas *sweeps* the gas through the interior of the system and it obeys the laws of gases. Any gases in the system equalize pressures so that vapor pressure equals liquid pressure. The dense gas can carry a lot of moisture vapor that would not be picked up by the light gas of a vacuum. Refrigerant vapor carries a lot of air with it.

I used this triple evacuation method on over a hundred residential hermetic condensing unit installations in five years and never had a burnout on a one of them. That was in the 1950's and to do that good after more than ten years gives me a lot of confidence in the procedure.

Think when you *sweep* the system. If you are pulling a vacuum from the suction line, introduce the dry gas from some other place in the system. If there is a gauge connection on the discharge valve, introducing dry gas at this point will sweep everything from that point to the vacuum pump connection. You may not be able to do this if you are keeping the compressor valved off during evacuation. In that case see if you can connect to the receiver inlet valve or the liquid valve. If no other connections are possible, consider soldering or brazing a Schrader valve in the line to get as much sweep as possible.

Think when you hook up your vacuum pump. The lighter the density of the gas you are pumping, the more line friction loss you will get. At 29 inches of vacuum, a quarter-inch tube will carry very little gas. Use the largest size line possible. Use the shortest line length possible. Are you pulling a vacuum on the entire system or only back to the expansion valve? Expansion valves are normally closed unless you have superheat on the bulb. Are there any solenoid valves in the lines? They must be opened: either with the manual opening stem or electrically.

You can pull a vacuum from two points at once. Connect the pump with a valved manifold, if necessary, so you can change pumping directions without loss of vacuum. If you find it necessary to remove the expansion valve to do a good job of evacuation, install a jumper fitting in place of the valve and then reinstall the valve after you have finished the evacuation and pressurized the system to about 5 pounds gauge. Pressure of the refrigerant leaking out during reinstallation of the valve will prevent entry of air and moisture.

How much vacuum do you need to pull? Take a look at the properties of water table. If you pull down to 11.1 inches mercury absolute and the temperature of the metal in the system is 50 degrees F, you have not vaporized any moisture. You must pull down below the metal temperature so that heat is available for vaporization. Your properties of water table is your best guide as to evacuation pressures that may be necessary.

Measuring vacuum accurately is necessary if you are to be sure of your evacuation job. There are thermocouple-type vacuum gauges that are very accurate. The McLeod vacuum gauge is very accurate. It is possible to use a wet bulb thermometer in a test tube that is tied into the pump suction line and read the temperature corresponding to the pressure. I prefer not to use this wet bulb gauge as you can only install it in the pump suction line and use it for short periods. It does add moisture to the line. For years I have used a Meriam 5-inch U-tube manometer with metric scale. It is not too expensive and has proven to be accurate enough for all my work. It is ruggedly constructed and can be closed off when not in use to prevent loss of mercury. Connect your vacuum gauge to the system at a point remote from the pump connection. You want to know the pressure in the system, not in the pump suction line.

Watch the oil in your vacuum pump. If it turns milky, you are mixing a lot of the water vapor with the oil as the vapor condenses.

Oil is cheap but vacuum pumps are not. Change the oil as often as necessary. Be careful if you are evacuating a system that has been flushed out with solvent. R-11 and trichlorethylene are solvents that are commonly used for this purpose. Even if they are blown out with nitrogen, the remaining solvent in the system will vaporize and then condense in the vacuum pump. The only way you can detect it is by a rise in the oil level in the pump. If this happens, stop the pump and draw a sample of the oil. If it is thinning out, change the oil.

These solvents are highly volatile and take an awful long time to pull out of a system. R-11 is the least damaging to a vacuum pump. If you must work on a system that has been flushed with other solvents, buy a case or two of vacuum pump oil and have it available on the job. For the same reason, never flush out or clean a vacuum pump with solvents. You may spend several hours pumping solvent vapor out of the pump instead of pulling a vacuum on the system.

While you are evacuating a system, look it over carefully. Feel the piping, the bottom of the receiver, low spots in the piping. Cold spots will tell you if moisture is boiling out at these points. A few strategically placed heat lamps will shorten evacuation time and enable you to do a much better job in some instances. **Think** about it.

The single evacuation procedure is sometimes specified in the installation and start-up contract. The serviceman is then looking for an indication of when the evacuation procedure is completed. Recommended procedure is to stop the pump and close the line valve when pressure in the system is close to the freezing point of water. Note the pressure and allow the system to stand for at least one hour. Check the pressure to see how much it has risen in this period. **If there are no leaks,** any increase in pressure will be due to water that has evaporated. No increase in pressure indicates that all water that can evaporate has been removed from the system. It usually requires more than one pull down to do the job.

DRIERS AND STRAINERS

Regardless of how well a system is evacuated and cleaned, driers and strainers should be installed in the lines. In the early days of refrigeration and air conditioning nearly all compressors had a suction line screen installed in them. Liquid line strainers were supplied with the compressor. The problem of moisture was handled with a temporary drier that had to be removed from the lines before it could powder and be carried through the system. Some manufacturers recommended the addition of small amounts of methyl alcohol to the system to mix with the water as an antifreeze. This was not true of all manufacturers. Some of the early machines were painted on the inside surfaces with a Bakelite paint to reduce refrigerant losses through the pores of the castings. Alcohol is a solvent of Bakelite and the soft paint was washed through the system plugging valves and strainers.

Better casting metals did away with the need for this interior painting and the advent of hermetic compressors has ruled out the use of alcohols. Motor winding insulation is not always safe against alcohol attack. I still see references in some installation and service manuals to the use of alcohol in small amounts in the system. The serviceman should be extremely careful about this. The manufacturer of the condensing unit you are installing may not be the manufacturer of the compressor. Check with the local agent for the compressor manufacturer before you follow such a recommendation.

Instructions, warnings, and recommendations of the original equipment manufacturer have a way of getting lost or forgotten by the time the equipment passes through the unit manufacturer's hands and gets to you. I recall one make of bolted semihermetic compressors whose manufacturer put out instructions never to tape or solder the liquid and suction lines together for liquid line subcooling. Too warm suction gas did not cool the hermetic compressor enough. This resulted in high head temperatures that caused the head to warp and blow out the gasket under the partition rib between suction and discharge cavities. This same compressor was sold to another manufacturer of condensing units and their installation instructions strongly recommended soldering liquid and suction lines together. The failure to note and pass on the original equipment manufacturer's warning resulted in a lot of service calls. Customers and dealers were both unhappy in this case.

The hermetic compressors require much cleaner and drier systems than before and this brought on more and better driers and strainers. Desiccants changed to types that could be permanently installed without the danger that they would powder and clog openings. The combination of driers with good built-in strainers made one unit all that was necessary. Molecular sieve and activated alumina driers do not dissolve in water as the old calcium chloride driers did. Molecular sieves trap moisture particles in and on their surfaces. Activated alumina and silica gel desiccants adsorb moisture on their surfaces. Particle structure is such that these materials present extremely large surface areas for this purpose. There is no reason not to install a drier permanently in the system anymore.

Drier installation practice for many years called for installing the drier in a bypass so it could be used only when moisture trouble was experienced. This is a hangover from the days of the dissolving material driers. Today's driers should be installed in the line as full-flow driers. A moisture indicator on the downstream side of the drier will indicate when the drier needs to be replaced. The drier should have shutoff valves on both sides if the system is very large, or large refrigerant losses could be expected while changing the drier or the drier cores. A bypass around the drier can be installed if continuous operation is necessary.

The main argument I have heard against permanent installation of liquid line driers is that they might restrict the line and cause a pressure drop that would cause flash gas to form in the line. A properly sized drier will not do this. Today's molded block driers are also exceptionally good line filters and they can and do restrict line flow if they fill up with contaminants. When this does happen, the drier has served its purpose, and should be replaced.

A moisture indicating sight glass downstream from the drier may or may not show bubbles with a restricted drier. If it does show bubbles, remember that this may mean a restricted drier or it may mean a refrigerant shortage. You can check for drier restriction by means of temperature. Any restriction will reduce the pressure of the refrigerant leaving the drier and this in turn will reduce the temperature. It's the law of gases again. A sight glass upstream of the drier will not serve as an indicator of drier restriction, but will definitely indicate refrigerant shortage if it shows bubbles. Regardless of where you install the sight glass, make it a habit to check the temperature on both sides of the drier when you are on the job. Either location of the sight glass still means you have to **think** about the drier.

There are times when I install two sight glasses on the liquid line. A standard sight glass upstream from the drier and a moisture indicating sight glass downstream from the drier. This has become almost standard procedure for me when I am installing a replacement compressor and driers after a burnout. Refrigerant is an excellent cleaning fluid and will wash the fine carbon particles from the piping in the system. Most driers will remove a percentage of these carbon particles. It takes a few drier changes to get them all out. Two sight glasses will tell you how effective the driers are in removing carbon.

The serviceman should be aware of the effect temperature has on an adsorbing-type drier. The colder the drier is, the more moisture it will adsorb. If the drier is holding all the moisture it can in cold weather, it will release some of this moisture when warm weather comes. Thus a system that indicates dry during the winter and changes to wet when warm weather comes means that the drier should be changed.

Permanent liquid line driers are fairly small when it comes to moisture holding capacity. A 33 cubic inch Sporlan Catch-All is the size usually used on a 3 or 5-ton residential central unit. This has a capacity of 152 drops of water at 75 degrees F or 122 drops at 125 degrees F. You might have to replace this drier once or twice if you did not get a good job of dehydrating the system with evacuation. I usually install a temporary replaceable core drier when I first start up the unit, Fig. 12-1. I can put a high capacity core in this drier that will hold 445 drops of water at 75 degrees F or 345 drops at 125 degrees F. It is a lot less expensive to buy and use the replaceable cores than it is to change the throwaway driers and the extra holding capacity means you use a lot less of them. Make the connection for your permanent drier and use adaptors to pipe your service drier in its place. When you have the system cleaned up and dry, you can remove the service drier and install the permanent drier. Use two sight glasses when you install the service drier and take them off when you are through.

FIG. 12-1 Temporary drier.

My personal preference in driers is the activated alumina desiccant. Activated alumina is not only a good desiccant, but it is also an exceptionally good substance for removing acid. Any acidity that develops in a system over a period of time will be circulated through the system by the refrigerant and the oil. Since all the refrigerant and oil in the system will pass through the drier eventually, the activated alumina will keep the acid number of the refrigerant and oil at a low level.

The sharp odor of refrigerant we used to have to contend with on the R-12 systems of the 1930's is something that I have never found on today's systems equipped with driers. The prevention of acid formation in today's systems is one reason for their longer life.

13

Soldering, Brazing and Leaks

Soldering and brazing are skills that every serviceman should have to be proficient in his work. As in every skill, a knowledge and understanding of what takes place during the process will help the serviceman to do a good job. Soldering and brazing differ from welding in that the metals to be joined in the process are never melted. A filler metal is melted in the joint and bonds itself to the metals to fill the gap and provide a strong force that holds the joined metals together. Filler metals with melting points below 800 degrees F are known as solders. Filler metals that melt above 800 degrees F are known as brazes. Whether the serviceman solders or brazes a joint depends on the melting temperature of the filler metal he uses in the process. When the metals to be joined are actually melted in the process, the procedure is called welding. A filler metal with the same melting temperature may be used in welding if it is necessary to fill up space between the joined metals.

Filler metals used in soldering and brazing are almost always alloys. Different alloys are used for different results. The plumber and the cable splicer must use solder alloys that melt and then freeze over a wide temperature range. An alloy that will do this can have both liquid and solid particles at the same time, Fig. 13-1. This results in a mushy mixture that allows them to wipe a lead joint for joining lead-covered cables or lead pipe to brass fittings. A very good wiping alloy is 38 percent tin and 62 percent lead.

An alloy which melts and freezes at a single temperature is known as an eutectic alloy. They are available for the special purposes that call for them.

The most common alloys for joining copper tubing have been 50/50 solder, 50 percent lead and 50 percent tin; and 95/5, 95 percent tin and 5 percent antimony. 50/50 solder was the first to be used in the refrigeration field and is still widely used for water piping and low-pressure refrigerant work. 95/5 solder has a higher melting point and greater holding strength and is recommended for refrigerant pressures found in R-22, 502 work.

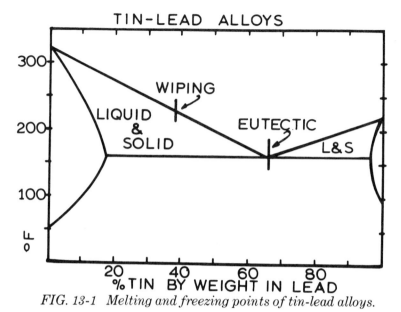

FIG. 13-1 Melting and freezing points of tin-lead alloys.

Silver bearing solder is a fairly recent addition to the soldering alloys. The addition of silver means a higher melting point and much more strength. Solders bond to clean metal both by adhesion and some alloying with the metal. It is molecular attraction that causes this adhesion and alloying. Tin has better properties of adhesion and alloying than lead. Silver is better than tin. Silver also has the ability to bond with steel that tin and lead do not have.

The requirements of the job dictate what kind of solder will be used. For electrical connections and noncorrosive coverings, a 40 percent tin, 60 percent lead alloy is all that is needed and is also the cheapest to buy. 50/50 solder is usually alittle cheaper than 95/5. Silver bearing solders are about twice as expensive as 95/5. The serviceman must select the proper solder alloy for the requirements of the particular job he is doing.

Solder will flow because of capillary action. Heat solder on a cleaned and fluxed metal surface and you will see the flow of the alloy as you heat the metal. Confine the alloy between two metals with a small space separating them and you can flow this metal alloy as much as 22 inches by capillary action. Solder alloy in itself does not have sufficient strength to withstand much pressure. Confine the solder alloy in a small space and you increase the strength because it must shear for quite a distance in respect to its thickness before it lets go.

This is the reason that solder fittings are male and female with some length to them. They are close fitting. The space between tube and fitting is usually from .002 inch to .005 inch. If the space is too small, solder will not be able to flow and cover all the surface of the metal. If the space is too large, capillary action will be cut down and the joint strength will be lessened because the softer alloy metal will have more room to shear. Copper fittings are designed and manufactured to the proper tolerances and lengths to give the maximum holding strength possible, Fig. 13-2. You will sometimes hear these referred to as *sweat* fittings. The action of flowing soldering and brazing alloys by applying heat to them is known as *sweating*.

Two lengths of copper tubing can be joined by using a sweat coupling. The serviceman can also use a swedging tool to expand the end of one length of tubing to fit over the other tube. This eliminates a fitting and one solder joint. A wide variety of sweat fittings are available. Fitting material may be wrought copper or brass alloys. All fittings are sized to fit with close tolerances to the copper tubing used.

Sizing nomenclature may be puzzling to the serviceman. Most refrigeration and air conditioning men refer to tubing and fitting sizes by the outside diameter of the tubing. If you are to run a liquid and suction line for a residential air conditioning unit, the specs may call for ⅜-inch OD liquid line and ⅞-inch OD suction line. When you go to the supply house to purchase this copper tube, be sure you specify that these are OD sizes. If you are buying from a plumbing supply house, you have entered another world. Plumbing nomenclature is entirely different. They use what is called nominal size instead of OD size. ⅜ OD is ¼ nominal. ⅜-in. nominal is ½-in. OD. The plumber's ½-in. line is the serviceman's ⅝-in. OD tubing. You have to think when you buy tubing, and you had better keep on thinking when you buy the sweat fittings to connect your tubing. It is a frustrating feeling to get miles out in the country and find you have ½-in. ID fittings to connect ⅝-in. OD tube.

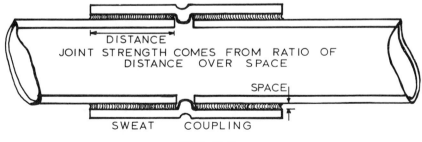

FIG. 13-2

This confusing state of tube sizing comes about because the iron pipe used in plumbing and pipe fitting is sized by the inside dimension. Standard weight ¾-in. pipe has an inside dimension of ¾-inch. Pipe threading dies are sized to fit the outside dimension of this pipe. Extra heavy pipe (schedule 80) has the same outside dimension so that the same threading dies may be used. The inside dimension is smaller.

¾-inch nominal copper tube is the refrigeration man's ⅞-in. OD tube. The outside dimension is constant so that all ⅞-in. ID fittings will fit it. The inside dimension of the tube varies depending on the wall thickness. Refrigeration tube is usually packaged in sealed coils for soft tubing and in capped 20-foot lengths for hard drawn tubing. The interior surfaces have been deoxidized and the tubing dehydrated and all air displaced with dry nitrogen before sealing the tube. Wall thickness of tubing varies from .030 for ¼-in. OD to .050 for 1-⅛ OD. This is type L or ACR tubing. Water and gas tubing is usually type K. It has a thicker wall and is not cleaned or capped. The serviceman will do well to stay with the cleaned and sealed ACR tubing in his work. Whenever he uses part of a roll or length of tubing, he should reseal the ends of the unused tube. Cleanliness of the interior of the system is too important to neglect this precaution.

Cutting tubing for soldering or brazing was originally done with a hacksaw. A fine-tooth blade, 32 teeth per inch, is necessary to avoid hanging up in the thin wall. The cut should be square so that the tube will contact the fitting shoulder to prevent solder loss into the tubing. I still have a Beaver square-end sawing vise from the old days.

Most tubing cutting today is done with wheeled cutters. Properly used, these wheel cutters make a square cut. If the wheel shaft or bore gets worn, the wheel will not track and will travel along the tube as it rotates. When this happens, your tubing cutter needs replacement parts. Starting the cut with too much pressure or with a dull, worn cutting wheel will result in a deformed tube that will not fit the inside of the fitting. **Think** when you are cutting tubing.

Whether a saw or a cutting wheel is used, you will have burrs on the cut tube end. This burr should be removed. Single and multiple blade reamers are available for this. A stout knife or a mill file can be used. The idea is to remove the burr and leave a square end on the tube. Chamfering is not necessary. Regardless of how you remove the burr, make sure that the metal taken off is not left in the tube. Hold the tube with the opening down and tap the tube to make sure all the metal particles fall out.

If you are working with installed lines and cannot hold the tubing for particle fallout, make a clean rag plug held with a twisted piece of coat hanger wire. Insert this in the tube and remove it slowly after you have cleaned the burr to pull all the metal particles out. Your care in doing this may save the customer many dollars of service work later. If the customer questions your care in doing your work, you should tell him your reasons for doing it.

CLEANING

Solder will not bond with metal surfaces unless they and the solder are clean. By clean, we mean free from oxide and other films. Copper and brass alloys oxidize readily when exposed to the atmosphere. This surface oxidation forms a protective film over the metal that stops further corrosion. It must be removed to allow the solder to bond tightly with the metal. Any oil, water, or grease on the metal will interfere with the solder bond.

Cleaning may be done by applying chemically active fluxes to the surface. This is a fairly common practice in plumbing work. These active chemicals are very damaging when they enter into a refrigeration system and it is almost impossible to keep them out of the interior of the tubing when they are used.

Standard practice is to clean the tubing and fittings with specially shaped wire brushes, or sandcloth, or steel wool or abrasive pads. Surface cleanliness is the important item. Surface roughness is not necessary. Oxide and metal particles removed during the cleaning process must be kept out of the system. Hold the tubing or fitting so particles will fall out. Tap the tube or fitting to help dislodge particles.

Do not blow on a cleaned surface, The moisture from your breath will form a wet film on the metal. **Do not touch a cleaned surface,** The acidic moisture from your skin will form a film on the metal that will prevent solder bonding.

FLUXES

As soon as you have cleaned the metal surfaces, give them a thin coat of flux. All that is needed is coverage that will prevent air from contacting the surface. Any more than this will enter into the system and could cause damage. Noncorrosive fluxes are available at your wholesaler. Most of them are in paste form and can be applied to the metal with small flux brushes. I have found that cotton swabs, available from a drug store, do a good job of covering the metal surface without leaving excess flux on the surface. They are cheap and should be discarded whenever they become dirty or worn.

Flux is not an expensive item. Brushes and swabs sell for a few pennies. It does not make sense to scrimp on these items and take a chance on a leaking joint that may take hours to repair. Keep flux covered when not using it. If there is any chance that it may have become contaminated, throw it away. Many small cans of flux are better than a large jar of it. Think about it.

Heating a solder joint is usually done with a torch. The serviceman may use an acetylene-air torch such as the Prest-O-Lite or he may use an LP gas torch. Soldering can be done with an oxyacetylene torch if the flame is adjusted for the work. A neutral flame with size to cover and flow around the fitting is necessary. We are trying to heat the fitting and the tubing all over so that all of the metal reaches the soldering temperature, as nearly as possible, at the same time.

Move the flame around to heat the fitting and the tubing at the same time, Fig. 13-3. Touch the solder wire to the joint crevice at a point where it is out of the flame to see if the metal is hot enough to melt the solder. When the solder starts to flow, move the flame to the other end of the fitting until you have used up the amount of solder needed to fill the joint. Then remove the flame at once. Eighth-inch wire solder requires an inch in length per inch of fitting diameter to fill a joint. A half inch of solder is enough to fill a ½-inch OD joint.

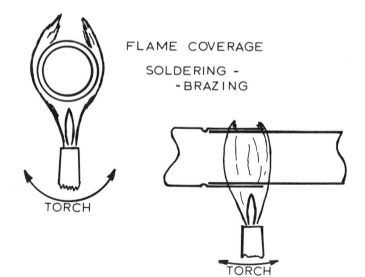

FLAME COVERAGE

SOLDERING -
-BRAZING

TORCH

TORCH

FIG. 13-3

You may continue to pass the solder wire around the outside of the joint until it cools, providing the excess solder is running off **outside the joint.** If you keep flowing solder into the joint, and it does not run off outside, then it has to be running into the inside of the tubing. Solder beads and drops inside the system are potentially damaging items. Usual practice is to let the joint cool a few seconds and then run a small bead over the joint crevice. This joint bead is not necessary and does not add any strength to the joint. A much better practice is to wipe the joint crevice area with a rag. You should be able to see a continuous ring of solder in the crevice after the joint has cooled. Excess flux on the joint should be wiped away or washed off. Any flux will eventually corrode the tubing if left on long enough.

The secret of soldering is to get the fitting just hot enough to flow the solder completely through the joint, but no hotter. Excess heat will burn off the flux and oxidize the metal before it can be covered by solder. Excess heating is the cause of more leaking solder joints than not enough heat. Insufficient or improper cleaning of tube and fitting is the biggest cause of leaks. Even dirty solder can cause leaks. If you keep your roll of solder in the tool box where it can become covered with oil or dirt, you may transfer this to the joint when you apply the solder. Wipe the solder off before you use it. I usually clean the section of solder with sand cloth or steel wool before I use it. I also bend a short length at a right angle to help me judge how much solder I am feeding into the joint.

Ever wonder why Prest-O-Lite comes in B and MC tanks? This goes back a long way. It was originally put out as the newest thing in automobile headlights. No more kerosene to pour in and no more wicks to trim. . . instant lighting carbide headlamps. The B tank was designed to strap on the running board of the Buick. The MC tanks were for motorcycles.

REPAIRING LEAKS

When you find a joint leaking, either on a freshly made solder joint or an old one, don't waste time heating the joint and trying to pour more solder in it. The only way to fix a leak is to take it apart and do it over. If the leak is due to insufficient cleaning, you will not get solder to bond until you get the metal clean. If the leak is due to overheating, the metal must be cleaned again before solder will bond. If the leak is due to distortion or pressure between tube and fitting that would not allow solder to flow into the space, the solder flux was driven off when it was heated and you must clean and flux again before you can get solder to bond. If this is a leak in an operating system, refrigerant leaking here has carried oil with it to contaminate the joint. You must get it clean before you can get solder to bond.

Heat the joint just enough to get it apart. Do it slowly and carefully, as this is an easy way to overheat a fitting. Be sure to heat the joint uniformly. All of the solder has to be liquid all the way around the joint before it will let go. Once you have the joint apart, inspect it carefully to see if you can tell why it leaked. Knowing what caused the leak is half the battle in repairing it.

Tinning the joint is often a great help in repairing leaks. Clean the tubing and flux it. Heat it and coat it with solder, wiping off the excess with a clean rag. Do the same thing to the fitting, wiping off excess solder with a clean rag swab on a twisted wire. You can clean and flux the tube and fitting in the regular way so that they will fit together with the thin solder coating on them. This solder coating or tinning will help in getting a full flow of solder in the joint.

Oxide films formed inside the tubing during soldering are tight on the metal surface and will not flake off into the system. Excess flux running into the tube during soldering is damaging and need not occur if the serviceman is careful. Loose solder in the system is the sign of the lazy, sloppy serviceman. Heat damage to valves, sight glasses, and controls with solder end fittings are the sign of the serviceman who does not think or read. These items are packed with an instruction tag that tells how to install them so they will not be damaged. Look for this tag and read it carefully. If you are in doubt about heat damaging an item, wrap it tightly with a water-soaked rag. This water creates a heat sink that can absorb a lot of heat while keeping the metal below water's boiling point.

When you get down to soldering the last joint in the system, you may find you are having trouble with it. This is because you are now working on a closed system. The air inside the tubes is being heated and expands in accordance with the laws of gases. This creates pressure and it tries to blow out of the joint that you are trying to fill with solder. Open a flare union or flange union somewhere that will relieve this pressure. Even if you get a joint to accept solder and stop blowing, the minute you take the heat away, the air inside cools and pulls air in because of the lowered pressure.

If you are trying to repair a leak on a joint that has water or oil inside it, the metal will never get hot enough to melt the solder until all the liquid is vaporized. This could create dangerous pressures in a closed system. You may find that the quickest solution to a problem of this nature is to cut the fitting in two with a hacksaw and drain the tubing. You can then remove the pieces and install a new fitting.

Always see that you have adequate ventilation when repairing leaks. Fluorinated hydrocarbons break down into toxic gases when you pass them through a torch. Check the ventilation before you start. Don't wait until you get sick to do it.

BRAZING

Brazed joints require temperatures much higher than soldering. 45 percent silver alloys flow in the 975 to 1000 degree temperature range. The 6 and 15 percent phosphorus alloys flow at 1250 to 1300 degree temperatures. These are close to the melting points of copper tube and fittings. Some brass alloys used in valves cannot be brazed with the higher temperature alloys because the brass melts first. Read the literature that comes with these sweat fitting valves before you braze them.

Cleaning joints that are to be brazed can usually start an argument among servicemen. The phosphor copper alloys can be used to braze tubing and wrought copper fittings without previous cleaning. You will find that you do the job better and quicker if you clean the joint first. If you are working with copper tube and brass fittings, cleaning and fluxing will make the difference between a tight joint and a leaking joint. It is a lot less expensive to properly clean and flux this joint before you start than it is to replace the fitting after you have tried to braze it and failed.

The flux most often used in brazing is a boric acid paste. It is active enough to clean oxide coatings but will not remove oil and grease coatings unless you use an awful lot of it. This means you run the risk of putting boric acid inside the system. Clean the joint just as you would for soldering and put only a thin coat of the flux on the metal. It is best to coat only the male section and let it carry flux to cover the female section of the joint. This keeps excess flux out of the system.

Brazing copper to steel requires the 45 percent silver alloy. This is expensive but the ability of silver to bond to steel means you get a very strong joint with very little brazing rod. The lower bonding temperature of the 45 percent alloy means that you will get much less oxidizing on the inside of the tube. The higher temperatures of the other alloys create a thick oxide film in the tubing that comes off and contaminates the system. To prevent this happening, dry nitrogen should be fed through the tubing to remove all oxygen carrying air from the inside. A needle valve regulator set for a flow of 20 cubic feet per hour will be enough to do the job. You can get these special regulators from most welding supply houses. You may find that using the lower temperature 45 percent silver alloy is cheaper than the nitrogen setup.

The torch flame for brazing is the same as for soldering. It may be larger because we need more heat, but the idea is to cover the joint and bring it all to the brazing temperature at the same time. Brazing alloy flows just like solder. It fills the joint just like solder. Any alloy outside the joint is wasted. The strength of the joint is only in the alloy that is between the metal walls of the joint. Once you have filled the joint space, the job is done. If you were able to flow one inch of eighth-inch square rod into the one inch joint, the joint is filled. The smaller rod requires proportionally more length. Two inches per inch diameter on the eighth by sixteenth flat rod and about three inches of the sixteenth round wire rod. You can dab on extra metal outside the joint by manipulating the heat, but you will not add to the strength of the joint by doing so.

Making good soldered or brazed joints requires attention to the fit of the tube and fitting. If the end of the tubing that is inserted into the fitting is out of round or has flat spots or dimples in it, you will not get alloy bonding in these spots. When you clean the tubing, these spots will show up as dark areas that have not been cleaned. It is possible to round the tube by inserting just the first part of a swedging tool in the tube. I have some special rounding tools that were made for me by a machine shop to round tubing up to three inches OD. It is almost impossible to remove these dents in tubing with a hammer and not further deform the tube. On the average job, the serviceman will be able to make better joints if he cuts back the tube to remove out-of-round conditions.

Don't clean a joint and then put it together without fluxing it. I was sent out one time to silver braze an installation for a dealer whose men did not have the facilities to do the job. When I arrived, I found that they had all the tubing and fittings cut to length and hung in place. All joints had been cleaned. Clean tubing will oxidize if it is not covered with flux. When two clean surfaces are placed together and then allowed to oxidize, the oxidation results in a rust weld of the metals. Rust welds are familiar things on steel but you may not realize that the same thing happens with copper. I had to open up the joints to flux them and had a lot of trouble doing this because of the oxidation welding. Some of the joints came apart after I heated them but quite a few were so bad that I had to cut them out and install new fittings.

When you are working with the torch, you need some kind of a guide as to when the metal is hot enough for alloy to flow. Brazing and soldering flux can be seen running from the joint as the heat builds up. Once this flow of flux stops, you are close to melting temperature. I usually try to melt a drop of solder on the top of the crevice. When the drop flows into the crevice, I know the metal is ready to flow solder as it is applied. Work quickly when you reach solder flow temperature and you will have less chance of overheating the joint.

Brazing flux dries when heat is first applied and is liquified by the heat slightly below alloy melting temperature. Watch for this liquefying of flux and be prepared to start feeding rod to the joint. You can do the same trick of placing a drop of alloy on the crevice with brazing alloy as you can with solder alloy. As with any skill, practice makes for more skill. Try practicing on soldered and brazed joints with scrap tubing in the shop. What you learn in the shop will come in handy when you are squeezed in a spot trying to make a tight, leak-proof joint on the job.

Watch that flame, The amount of heat given off by a torch can set fire to inflammable materials close to it. Look around before you light up. If you **think** that there is any danger of fire, put up some kind of flame barrier to prevent fire from starting.

LEAK DETECTION

When a refrigerant piping installation is completed, the next step is to test it for leaks. The initial test is usually to pressurize the piping with either refrigerant or an inert gas to a low pressure, say 30 psi. If you don't hear it blowing, you watch the gauge to see if the pressure remains steady or leaks off. Sometimes you can be fooled by this. With the condensing unit valved off, pressure added in either the suction or the liquid line will take some time to equalize through the TXV to the other line. Think about this when you start testing for leaks.

The next step should be to pressurize the system to at least 150 psi with dry nitrogen and run a soapsuds test. If no leaks are found, the pressure can be blown out and refrigerant added to bring the pressure up to 30 psi and then nitrogen added to bring the pressure up to 150 psi. Then you can test with a halide torch. If no leaks are found, a final test can be made with an electronic leak detector.

The electronic leak detector is too sensitive to be used for tests where a large leak may be present. It can and will indicate a leak at a fitting when the actual leak is several joints away. A clean and well cared for halide torch is more sensitive than most servicemen realize. Properly used, it will find all but the most minute leaks. There are some complaints that the change in flame color is hard to see in a bright light. Make a shield for the torch out of a tin can with the inside painted black and you will be able to use the halide torch in bright sunlight. Cut a narrow slit on the side and bend the sides out for a viewing slot. Drill or punch small holes in the sides next to the top for an air outlet. Solder a wire to the can so that you can place the shield over the burner and hold the wire with the same hand that holds the torch handle.

How much pressure to use for leak testing will depend on the system and its components. As a general rule, exceed system operating pressure by at least 50 percent. The installation manuals that come with the unit generally give the proper testing pressures. More important, they give pressure maximums to prevent damage to the system components. The best thing you can do is to read the instructions. While we are on the subject of test pressures, I would like to point out that a vacuum test is not a leak test. When a serviceman tells me that he held a vacuum on the piping for a week without losing it, I wonder why anyone would want to brag that his piping work is so good that it will hold 15 psi air pressure from leaking in. Once upon a time (1937) I forgot to solder a 1-⅛in. ell in a suction line and the solder flux alone held the R-12 for almost three months.

When you are hunting for leaks after the system is operating, you have a little more trouble. You can bring discharge pressure up, to help in leak testing, by blocking the air or water from the condenser. It will not stay up too long so you have to work fast. Suction line pressures are hard to raise. On water chillers they can sometimes be raised by running hot water through the chiller. This is risky business and should be done with extreme care. The electronic leak detector is the best tool for leaks in this case.

Oil around a joint is usually a sign of a leak. Double check this by cleaning the joint with solvent and clean rags and then covering or wrapping the joint to make sure the oil is not coming from some exterior source. Wrapping a joint in plastic wrap and sealing the seam and edges with plastic tape, to make it airtight, will sometimes prove out a leak. Use a lightweight household wrap and mold it to the pipe to exclude as much air as possible. If it bulges the next day, the pressure had to come from the joint. Building a dam around the joint and submerging it in water is sometimes possible. Mix soap with the water to reduce surface tension and make it easy for bubbles to form. Since the water will absorb some of the refrigerant, you should allow plenty of time for the test.

The red dye additives for leak detection have been cussed and discussed in service literature. I have used them three times for finding stubborn leaks. In all three cases they did the job and did not affect operation or damage the hermetic compressors in any way. They were very helpful in finding leaks caused by porous brass and iron castings in component parts of the system.

Leaks in condensers and chillers are about the hardest to find and pinpoint. If the water side of the unit is drained and valved off, leaks can sometimes be detected by checking at an opening at a low point in the system. They can be checked at a high point, but it will take longer for the refrigerant to fill the water space until it reaches the high level. If you find that you have a leak in a shell and tube vessel, you can determine whether the leak is in the tubes, and which tubes they are, by plugging them with corks. Buy enough corks from a wholesale drug firm to plug **both** ends of the tubes. Clean the tube ends with a brush until they are shiny and press the corks in firmly. Do not oil or grease the corks. Use water for a lubricant. After the corks are all in place, you can test the tube sheets for leaks around the tubes with a torch or electronic leak detector. Give the corks 24 hours or more, depending on the amount of refrigerant you have been losing, to *blow their top* and you will have pinpointed the leaking tube. Finding these troublesome leaks can be exasperating but they can be found with patience and **thinking.**

14

Lubrication

A refrigeration compressor is a machine with bearings and pistons. They require lubrication with a refrigeration oil that must, at the same time, be compatible with the refrigerants that are pumped by the compressor. Refrigeration oil must be able to operate at the high temperatures of the valve plate and must still function and not plug the working parts of the expansion valves at low temperatures.

Refrigeration oil is a special oil that is manufactured to meet the specifications of the compressor manufacturers. Most compressors come charged with the proper quantity and type of oil and the serviceman does not have to worry about it.

When the serviceman is required to add or change oil in a compressor, he must make sure that he uses the right oil. Oil is a manufactured product made from different types of petroleum bases. It has different types of additives in it.

I have learned from experience that mixing two different brands of oil can cause trouble. You can valve off a compressor and drain the oil from the crankcase. This will not remove the oil that is mixed with the refrigerant and coating the rest of the system. Make sure you use the right oil.

You will very seldom find the recommended oil mentioned by name in your service manuals. The exception is the case where the compressor manufacturer sells the oil under his own brand name. Other than this, most manuals will give a set of specifications for oil and say you can use oil that meets these specifications. The serviceman does not have the time to obtain oil specifications and compare them. Looking in a wholesaler's catalog, I find three brands of oil for sale. All three of them come in 150 and 300 viscosity.

One brand comes in regular and dual inhibited grades. Regular is for compressors operating under normal conditions. Dual inhibited is for compressors operating under high temperature conditions, stable to 348 degrees F. The wax floc points are minus 80 degrees F for 150 viscosity and minus 60 degrees F for 300 viscosity.

Another brand states that it complies with Federal Specification

VV-L-825a Type IV. The third brand gives prices only. All three of these are excellent oils. I have used all of them. I know which ones are preferred in different makes of compressors because I have checked with the local factory service managers and gotten the **word** from them.

There must be some government regulation that prevents manufacturers from writing in the brand names of oils they use in their machines. You will have to get your information in verbal form. If you cannot get it this way, try some other service shop that works on this make and see what they use. This is just one of the little headaches that bother the serviceman.

Oil must be clean and it must be dry to be used in a refrigeration system. If you cannot use all of the oil in the container on one job, the remainder must be sealed tightly if you expect to use it in another system. If this cannot be done, or you are not sure of the oil, do not use it. It could become a lot more expensive than buying new oil.

Oil does not wear out. As long as it is kept clean and not exposed to extreme high temperatures, it will lubricate indefinitely. The residential air conditioning unit and the household refrigerator will run for years and years without a change of oil. A dry system and a drier that will remove acid will help oil to last this long. The larger compressors subject oil to higher temperatures and oil is changed often in these units. The use of good liquid line driers, with acid neutralizing elements, in these large units will keep oil usable for several years. An oil test to determine the acid number of the oil should be made annually on these larger units.

The fluorinated hydrocarbon refrigerants have the ability to mix with oil and they carry a part of the oil charge through the system with them. When a system is first started, you can expect part of the oil to mix with the refrigerant and leave the crankcase. If the oil level gets too low in the crankcase sight glass, you will have to add oil to bring it back up to the operating level. You should only have to do this once. Before you add oil a second time, **think.** Oil does not leave the system unless it is lost with a refrigerant leak. Oil that is being carried away from the crankcase and trapped somewhere in the system is going to come back some day. When it does, it is going to be pulled into the compressor suction and pile up on top of the piston. Oil is a liquid and cannot be compressed. When you open up the crankcase of a 50-hp six-cylinder compressor and start scraping out the bits and pieces of the pistons, rods, cylinder sleeves, you find out what happens to a vapor compressor that tried to compress liquid.

If oil is being continually lost from the crankcase, check the system for traps or wrong piping practices that can cause oil to be held in the low side. Oil loss is something that is seldom found on the average small installation. It is not uncommon on larger installations. Capacity control installations can trap oil if they are not piped correctly. Study Figs. 14-1, 14-2, 14-3 and you will see how some of these conditions can be caused and how they can be corrected.

It is almost impossible to lubricate a compressor and not pump some oil past the pistons and into the discharge line. On smaller, medium temperature systems that is not a problem. On larger systems and low temperature installations, oil pumping is something that must be taken into account and provision made to correct the problem. Oil is carried along with hot discharge gas in the form of small droplets. Separation is accomplished by impingement. Direct the oil and gas mixture against a surface and the oil will remain on the surface and run down to a lower elevation while the gas is able to reverse its flow and leave at a higher opening.

Oil separators are commonly installed in discharge lines. Ammonia installations have oil drain legs and valves installed on all parts of the system. Ammonia and oil separate quickly and the oil goes to the bottom.

You will not find as much provision made for oil draining on fluo-

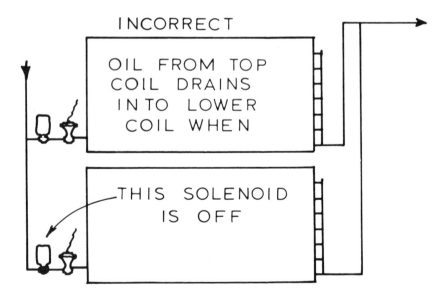

FIG. 14-1

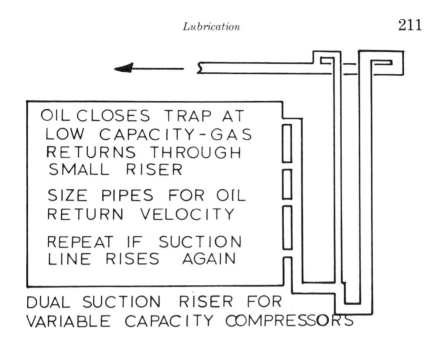

FIG. 14-2

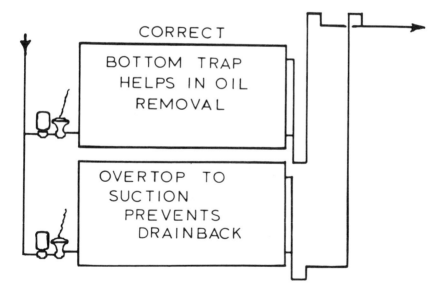

FIG. 14-3

rocarbon refrigerant systems. DX evaporators do carry a lot of oil back to the compressors in the suction lines but flooded systems are not as good at returning oil. I once diagnosed the trouble on a 250-ton flooded water chiller with a low side float control as too much oil in the chiller. Eleven 55 gallon drums of oil were drained out of the chiller. I cannot figure out how the operators could add that much oil to a system without thinking about where it was going and what effect it would have on operation.

Any evaporator that uses a refrigerant control to maintain a level of refrigerant liquid is a flooded evaporator. Finned tube evaporators and blower coils can be operated as flooded coils if a suction surge drum is added to them. Low side float valves are usually installed in the surge chamber. If several evaporators are in the system, a suction accumulator is added to catch any liquid that may slop over

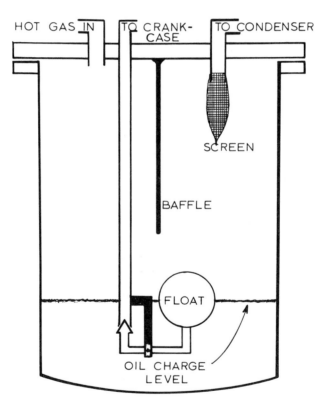

FIG. 14-4   Oil separator.

from the coil surge drums. Level switches in the accumulators cut off a liquid line solenoid when the level of liquid in the accumulators gets too high. The accumulator should have a liquid line coil in it and act as a heat exchanger also. Oil drain valves are usually installed at the bottom of accumulators.

Fluorocarbon refrigerants hold a lot of oil in suspension. Most of the oil is held in the upper level of the liquid. Oil in a suction line will seriously affect the operation of a thermal expansion valve by insulating the bulb from the refrigerant. Oil in a low side float chamber will lower the level of liquid in the evaporator and in turn raise the superheat. When the serviceman diagnoses his problem as oil in the evaporator, he is faced with the problem of how to remove it. The installation of a continuous oil removal system is the best answer. A heat exchanger can be used as an oil removal system if it is possible to pipe the liquid line through it. If this is not possible, an electrical heat element such as an insert type crankcase heater must be used.

The oil removal device is nothing more than a refrigerant still. Oil rich refrigerant is periodically taken from near the top of of the evaporator, and as close to the refrigerant control as possible, to partly fill the still. Heat vaporizes the refrigerant which is piped to the suction line. Oil remaining after the refrigerant is gone can be returned to the compressor or drawn off into a container. The amount of oil drawn off should be compared to oil added to the system to gauge the effectiveness of the still.

Crankcase heaters are required items on almost all compressors. If the compressor crankcase becomes the coldest part of the system on an off cycle, the refrigerant vapor in the system will follow the law of gases and migrate to the crankcase. Once there, it will condense back to a liquid and mix with the oil. Start up the compressor and this liquid will boil out of the crankcase and take all the oil with it. You will have to operate without lubrication until it returns. The chances are that the compressor will be badly damaged before the oil does return. A crankcase heater keeps the crankcase warm enough to assure that it never becomes the coldest part of the system and the oil is always warm enough to vaporize and drive off any refrigerant.

Some oil separators also have heaters installed on them. Otherwise, liquid refrigerant, that collected in the separator on the off cycle, would be dumped into the crankcase by the oil return float valve in the separator. You have to **think** about the location of the separator and the other parts of the system, and the possibility of temperature differences, to determine whether or not they need heaters. Even with insulation wrapped around the oil separator, the separator may cool off faster on an off cycle than the larger mass of metal elsewhere in the system.

The serviceman uses still other types of oil in his work. Cooling towers with gear reducer fan drives require special oils, usually turbine oils with special antioxidant inhibitors in them. Oil should be changed in the gear boxes every year due to the wet atmosphere in which they operate. Cooling tower operating manuals point this out and many include a long list of manufacturers' names and the brand name of the oil that is acceptable for use in the machinery. You will have to order this oil from your local oil distributor. Wholesalers of refrigeration and heating equipment do not carry it.

Oil used in motor and fan bearings should be carefully selected for the purpose. Don't just grab a can of motor oil at the filling station and use it for these bearings. Automotive oil has the widest variation of additives possible between makes and can create some extremely serious problems when different brands are mixed. On one service call, a large General Electric motor was shut off because the bearings were overheating. I was sent out on the call and checked the bearings. The filler pipe was full of oil, so I removed the bearing covers. I found the oil in the bearings had turned to jelly due to the mixing of different brands of oil. It was necessary to heat the bearing before the oil would drain out. After washing out with mineral spirits first and refrigeration oil second, the bearings were refilled with nondetergent oil and operated correctly.

Pick a good brand of nondetergent oil for use in motor and fan bearings and carry a supply with you in the proper viscosities. Keep it clean. Dirt can ruin bearings when added with the oil. A simple way to keep a quart of oil clean is to punch two small holes in the can. After filling your oil can, close these holes with sheet metal screws until you again need oil. If you will keep all your oil and oil cans in a metal-bottomed container in your truck, you will not have it running all over the bottom of the truck.

The serviceman working on larger equipment will also need grease and grease guns for grease-lubricated bearings. Here again, you have to know about greases. Chassis lubricating grease from the filling station will ruin a ball bearing. Ball bearing grease will not do the job in a water pump that requires lithium base grease to be effective in wet atmospheres. Keep your supply of ball bearing grease clean. Do not leave the container open longer than necessary to refill the gun. Use a separate gun for each type of grease you carry. Read the instructions on the bearing nameplate or the motor lubrication instruction plate. Follow these instructions. Above all, **think,** when you are lubricating bearings. More bearings are ruined from overlubricating than from underlubricating.

FLUSHING AFTER A BURNOUT

A burnout produces carbon and acids. The carbon is a visible product and can be seen in the refrigerant, in the oil, and on internal surfaces. The acids can be smelled and found with simple chemical tests.

The acid products will be present in the refrigerant and in the oil. **They can be very harmful to the serviceman.** Take every precaution possible when obtaining samples of refrigerant or oil, or discharging the system, to keep them from contacting your skin or spraying into the eyes. You can lose fingernails, eyesight, or a lot of skin if you are not careful.

An experienced nose will tell you if burnout acids are present. Acid in the oil, if strong enough, will bleach a red shop towel white. You can stir a sample of oil, or refrigerant oil mixture, in a small amount of water and then pour off the oil and check the water with the same paper you use for testing acid cleaning solutions. Any positive results from these simple tests indicate a fairly strong acid content in the system. The acid test kits available at wholesale houses are the most positive, and give a better indication of the degree of acid contamination.

The type and amount of cleanup procedure of the system when a burnout occurs depends solely on the serviceman's judgment as to how severe the burnout was and how much contamination is present in the system. One item on which to base your decision should be how easy is it to change the oil in the compressor. Oil is a natural scavenger of contaminents. If you cannot change oil easily in the exchange compressor, get the system as clean as possible before the exchange.

Most small systems with welded hermetics fall in the class where oil changing is very difficult. Flush these systems with pressurized R-11 solvent. Clean the expansion valves and take them out of the system, if possible, during flushing. Remove any strainers and clean or replace them before and after flushing. Remove liquid line driers before flushing. Do not replace them until you are ready to operate the new compressor. A sample of the flushing fluid can be tested to see how acidic it is after flushing.

The serviceman must use his own judgment as to when the system is clean enough and what type of strainers, filters, and driers should be installed in the lines to protect the exchange compressor. The fact that oil cannot be changed as easily as a suction line filter indicates that suction line filters are a must on this type of burnout exchange.

If the system has a compressor where oil changing is easy, it is usually a larger system that requires a lot of cleaning and flushing fluid. Let the amount of burnout products guide you as to whether or not flushing is needed. I drained oil from a 25-ton hermetic that was so acidic it ate away the surface of the concrete floor where some of it spilled. The discharge line had a flanged valve at the condenser inlet that allowed me to remove the line and inspect it without losing the system charge. The line showed no sign of carbon at the receiver end.

I decided that the system had shut down so quickly it had not carried contaminents beyond the compressor. We cleaned up the compressor and had the motor rewound in the factory service plant. It was reinstalled and a three-core drier installed in the liquid line. We changed the activated alumina drier cores four times and the oil three times as the sight glass and oil acid tests indicated were necessary. Expansion valves were cleaned twice. The second cleaning could have been dispensed with, but we had to do it to find this out. **Think.** Remember your past experience, use your own judgment as to burnout cleanup procedures. There is no set rule to go by. You must make the decision yourself.

15

Bearings

Sleeve bearings are commonly used in motors and fans. As the name implies, it is a metal sleeve that fits over the turning shaft. Clearance between the shaft and sleeve is originally set at about one-thousandth of an inch per inch of diameter, but increases with wear.

Bearing materials are usually brass alloy. Being a softer metal than the shaft, wear occurs mostly in the bearing. Lubrication keeps this wear to a minimum.

Lubricating oils tend to stick to metal surfaces. If you have tried to wipe up oil from a metal surface, you know how difficult it is, since oil spreads and resists wiping off. It takes many rags and sometimes a solvent to get all the oil off.

Oil is attracted to metal surfaces and capillary action draws the oil into the confined space between the shaft and bearing. A rotating shaft actually drags the oil with it, Fig. 15-1. When a shaft rotates at sufficient speed, the oil builds up, in the space between shaft and bearing, Fig. 15-2, into a uniform thickness and forces the shaft, into

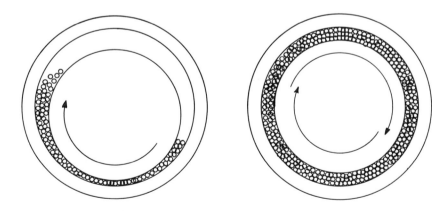

FIG. 15-1 Rotation begins. Shaft
　　　　drags oil film.

FIG. 15-2 Rotating shaft
　　　　pumps oil.

the center of the space. Metal-to-metal contact no longer exists because of the action of the oil.

Once the space between shaft and bearing becomes too great for this oil buildup to occur, bearing wear will increase rapidly. You may have noticed that, once a bearing starts to give any trouble, it deteriorates rapidly. It is all because of the excess space in the bearing. A belt that is too tight will also prevent the shaft from centering in the bearing. That is the reason why tight belts cause bearings to wear out so fast.

A continuously operated shaft will give longer bearing life than a shaft that starts and stops frequently. When a shaft is up to speed, it cannot wear the bearing because of the centering action of the oil film. Bearing wear occurs from the time the shaft starts turning until it picks up enough speed to drag the oil film into place and, on shutdown, from the time when speed drops too low to maintain the oil film until the shaft stops turning, Fig. 15-3. You can see from this that shaft speed is important to bearing lubrication. The next time you look at a propellor fan lazily turning in the breeze on an air-cooled condenser, **think about it.** Now you see why we have such a high rate of bearing failures on these fan motors.

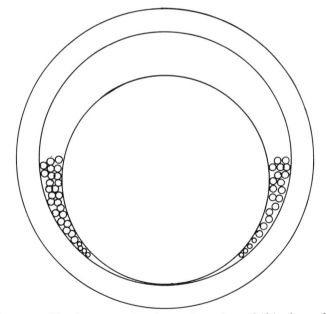

FIG. 15-3 Shaft at rest in sleeve bearing. Oil is forced out
          below shaft.

RING OILERS

Ring oiler type bearings have a square-walled opening in the top center and oil grooves on the inner surface. The top opening allows the ring to rest on the shaft and be turned by the shaft as it rotates, Fig. 15-4. The inner grooves distribute the oil, deposited on the shaft by the rotating ring, over the length of the bearing. Ring oiled bearings require fairly heavy viscosity oil and must be kept clean. It is common service practice to drain, flush, and refill this type of bearing at least once each year.

WICK OILERS

Wick oiled bearings usually have sloping walls on the top opening and oil distribution grooves, Fig. 15-5. Oil is carried from the reservoir to the top opening by capillary action of the wool yarn wick. Wool makes warm clothing because of the insulating value of the hollow wool fibers. These same hollow fibers carry a lot of oil up to the top of the bearing. Plug those fibers with dirt, or oil that is too heavy or gummy, or scorch the wool to close off the fibers, and a wool yarn wick is useless. If any of these troubles show up, the yarn wick must be removed, the reservoir cleaned, and a new yarn wick properly installed.

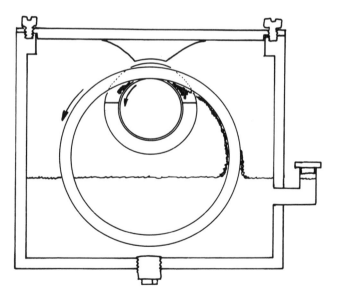

FIG. 15-4 Ring oiled bearing.

INNER SURFACE OF A BEARING-UNROLLED

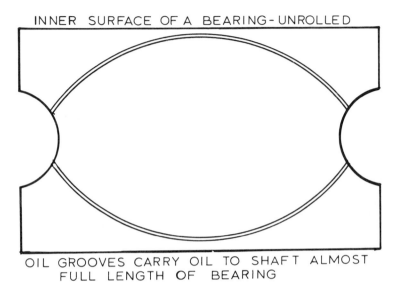

OIL GROOVES CARRY OIL TO SHAFT ALMOST
FULL LENGTH OF BEARING

FIG. 15-5

LIFETIME OILED BEARINGS

A lifetime oiled bearing usually has a felt wick that is fitted close-ly over the center or at one end of the bearing. The reservoir is large enough to hold a good supply of oil. The bearing housing and oil re-servoir is tightly enclosed to keep out dirt. The bearing itself may be of brass or copper alloy or it may be what is called a *sintered* bearing. Sintered bearings are made by compressing small metal particles together and then supplying enough heat to cause the particles to bond together without becoming a solid homogeneous mass. This results in a very porous metal structure. The inner bearing surface can be machined to form a uniform bearing surface. The porous met-al structure results in good capillary action that draws the oil through the metal. The same procedure is used in forming the mold-ed block cores of driers.

Sintered bearings are usually used only on small bearings that do not have to carry much weight. Larger bearings are usually made from brass bearing stock. The term *lifetime* does not refer to any specific length of time that the bearing will last. It means that the bearing has been supplied with sufficient oil to last it for its normal life. Life expectancy is determined by the operation of the bearing. The more often it stops and starts, the shorter its life. I have found some of the small lifetime bearings to have a screw in the bearing housing that is set in with thread dope. You can remove this screw and lubricate these bearings when necessary. Use lightweight clear oil such as Standard's *Fineoil,* or 3-in-1 oil for these small bearings.

BALL BEARINGS

The serviceman is finding more and more ball bearings used on equipment. They do a better job than sleeve bearings and will last longer if properly maintained. Improper or damaging maintenance and lubrication procedures cause more ball bearing failures than any other causes. This stems from lack of knowledge of ball bearings by the serviceman. Let's see if we can correct this.

Basically, a ball bearing consists of an outer ring with a ball race on its inner surface, an inner ring with a ball race on its outer surface, a number of balls between these rings and two metal retainers formed to hold the balls in the proper spacing between these rings. See Fig. 15-6. The important dimensions for a ball bearing are the width, bore and outside diameter.

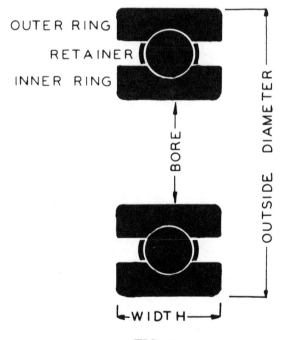

FIG. 15-6

Fig. 15-7 shows the relation of the inner and outer rings and balls in a Conrad type of bearing. This type of bearing is the most generally used. It has good radial load strength and good thrust load strength. The area of angular contact between the balls and the races is what determines the strength. As you can see in Fig. 15-6, the angular contact area is large. The Conrad bearing gets its name from the man who invented the method of assembling this bearing, Fig. 15-8.

Like most good inventions, it seems so very simple. With the inner ring off center, the balls are placed in the outer race and then the inner ring is centered, the balls spaced around it, and the retainers installed to hold this spacing. The serviceman can drill and cut the retainer on a failed bearing of this type to disassemble it for inspection to determine the cause of failure.

One other thing to be noted in Fig. 15-6 is the rounded bore corners and OD corners on the rings. This radius is necessary to fit a bearing in a housing or against a shaft shoulder. A rounded fillet is usually left on the shoulder for added strength. The radius clears these fillets and allows the face of the bearing to meet the housing or shaft face.

Fig. 15-9 shows a maximum capacity type bearing. The more balls there are in a bearing, the greater the load it will carry. In order to get more balls in a bearing, a loading slot is required. This loading slot reduces the angular contact area at the slot and this reduces the thrust capacity of this type of bearing. On a bearing with a loading slot, the thrust capacity is only 50 percent of the applied radial load: not the rated radial load, but 50 percent of the actual applied radial load. A retainer is used to maintain proper ball spacing.

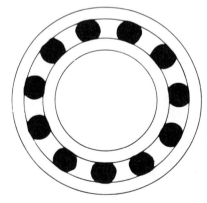

FIG. 15-7 Conrad type bearing.   FIG. 15-8 Conrad assembly.

Thrust type bearings are constructed as shown in Fig. 15-10. The outer ring is counterbored so that it has good angular contact on only one side. The ring is expanded by heating and slipped over the inner ring with the balls in place.

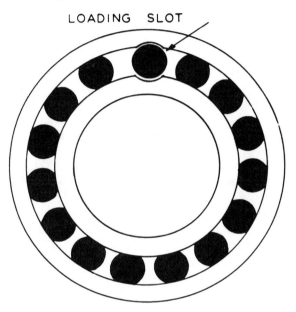

LOADING SLOT

FIG. 15-9   Maximum capacity type bearing.

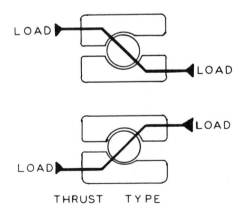

THRUST   TYPE

FIG. 15-10

As you can see from the force lines on the drawing, this type of bearing has excellent thrust capacity **in one direction only.** It also has the dubious distinction of being the most frequently mismounted type of bearing. Anytime you replace a bearing of this type, look at the face of it for the word THRUST on one of the rings. Load applied to this ring face will be transmitted through the balls to the opposite face of the other ring.

Fig. 15-11 illustrates two things: internal fit and preloading. Internal fit is also called radial play. It refers to the precision with which the balls are matched and is usually measured in ten thousandths of an inch. Some radial play is necessary to allow for heat expansion of the bearings and press fit stresses of housings and shafts. The amount of radial play determines the amount of end play in the bearing and the amount of angular contact area between the balls and races. The precision in machining shafts and bearing housings also determines the amount of radial play required.

Bearings are available in five grades from tight to extra loose. On a single row, light duty bearing with a one inch bore, this would be from minus .0001 to plus .0001 for tight to .0008 to .0012 for extra loose. Standard internal fit would be .0002 to .0005 in. Bearing dimensions are also graded as to precision of manufacture. Grade 1, for instance, has an allowable tolerance of .0006 in.; Grade 9 has .00015 in.

In some thrust applications, thrust bearings are mounted in duplex pairs. Two bearings, such as those shown in Fig. 15-10, may be mounted in the same housing with the bearing faces touching. They may be mounted back-to-back, face-to-face, or in tandem This not only gives greater thrust loading but it also reduces end play.

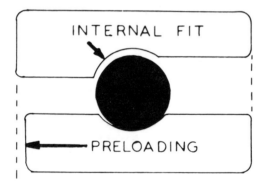

FIG. 15-11

Where greater control of end play is required, the bearings may be preloaded. Instead of both rings being ground to the same width (flush ground), one ring is ground slightly narrower. This reduces the internal fit and in turn reduces end play. Pre-loading is expressed in pounds. The amount ground from the ring is the same amount the ring would be deflected if it were supporting a load of equivalent weight. A word of caution to the serviceman, when replacing duplex pairs, either flush ground or preloaded, always replace **both** bearings. These bearings come in matched pairs just as V-belts are matched.

BEARING FAILURES

Every ball bearing has a given life span. This life span depends on two things: the load on the bearing and the speed at which it operates. If you double the bearing speed you reduce the life to one-half. If you double the load you reduce the life to one-eighth. Remember this the next time you tighten those V-belts. To see why this is so, look at Fig. 15-12. Regardless of the fact that the balls and races are of hardened steel, there is still some elasticity in the metal. Both the ball and the raceway are deformed by the load they are carrying. The ball flattens out where it touches the race and a swelling of the metal occurs on the leaving side of the ball. The race is depressed and a standing wave occurs in the metal just ahead of the ball and travels ahead of it as it rotates. Over a period of time, this constant flexing of the metal causes metal fatigue. The metal flakes off and is deposited

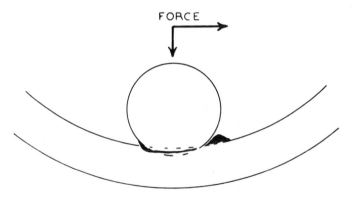

FORCE

ELASTIC DEFORMATION OF BALL
AND RACE BY LOAD AND MOTION

FIG. 15-12

in the outer raceway. Because of the rough surface, the bearing becomes noisy. This is normal in the average bearing after it has run some 8 to 10 thousand hours. If it occurs after only one or two thousand hours, look for some other cause: too tight belts or a bearing housing out of round.

Contamination is by far the largest cause of bearing failures. As opposed to flaking or spalling of the raceways, contamination causes scratching, pitting and scoring of the raceways. Bearing noise is usually intermittent when contamination first occurs. Rust forming in the raceways during the off season is a common form of contamination. With internal fit measured in ten thousandths of an inch, and dirt particles measured in thousandths of an inch, you can easily see why cleanliness is so important in installing and lubricating ball bearings.

Contamination prevention is simple. Keep everything clean. The bearing housing, your tools, your hands, the flushing solvent, the lubricant can all must be kept clean. Wrap the bearing in oil-proof paper before you lay it on the work bench next to the grinding wheel. *Don't* wash the bearing in solvent when you remove it from the package. The rust preventive coating on these bearings is compatible with all lubricants. Leave it on unless it has become contaminated. *Don't* spin the bearings with the air nozzle until they shriek like a banshee. The noise you hear is hard, brittle metal being tortured beyond its limits. You can lose a finger or two if it lets go and flies apart.

Brinelling is another cause of bearing failure. If the bearing is improperly supported during installation, the balls are forced against the shoulder of the race and tiny indentations are made in the shoulder of the race and in the balls. Operation with these indentations will cause spalling and pitting of the raceways. Indentations can also be made in the center of the raceway if the bearing is dropped. The rings may even fracture, in this case, because the brittle metal will not withstand a sharp shock load. To prevent this cause of failure, always use the proper mounting procedure. If you are pressing a shaft into a bearing, always support the center ring. If you are driving a bearing on a shaft, use a tubular drift on the inner ring and a rawhide hammer. Metal hammers can send fragments flying into the bearing. If you are pressing a bearing into a housing, press against the outer ring. If you are removing a bearing that is to be reinstalled, use the proper bearing puller and puller jaws to exert force on the proper ring.

Thrust failures in bearings are fairly easy to identify. If the balls are banded and the counterbored shoulder breaks down, then the thrust bearing is improperly installed. Check the direction of thrust against the mounting of the bearing. Excessive thrust on a maximum capacity bearing will show up as nicking and pitting of the balls and spalling of the races next to the loading slot. On a double row, internally self-aligning bearing, excessive thrust will result in spalling and flaking on one row of balls and one side of the outer ring race. On these two types of bearings, check to find the cause of this excess thrust.

Misalignment of a standard bearing will cause excessive retainer wear, possibly wearing through it and causing its loss. A misaligned bearing will show a very wide ball path on the rotating ring and a ball path from one side of the race to the other side in one-half its circumference and then back again, on the rotating ring. It can be caused by the shaft or housing being out of line or the shaft may be bent. The wide inner ring bearing shown in Fig. 15-13 can very easily be misaligned. The fact that the outer ring is supported in a pillow block with some flexibility will not prevent this type of alignment damage. This wide inner ring bearing is very common on commercial air conditioning units. The inner ring is locked to the shaft with either a set screw or an eccentric cam. If the clearance between the shaft and the bore exceeds .001 in., the inner ring will be cocked as shown. This causes opposite thrust shock loads at each revolution of the bearing, resulting in rapid bearing wear and, in some cases,

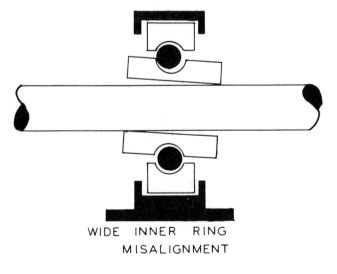

WIDE INNER RING
MISALIGNMENT

FIG. 15-13

breakage of the wide inner ring. The only remedy is to build the shaft up to the proper size.

Lubrication is essential to the operation of a ball bearing. There are only two points on a ball where a true rolling action occurs. In between these two points and on either side of them, the movement is a combination of roll and slide, Fig. 15-14. Rapid retainer wear will occur without lubrication because of the rubbing action be tween retainer and balls.

Without lubricants, the bearing heats and expands, clearances disappear and failure is very fast. Dirty lubricants have already been covered under contamination failures. Excess lubricant is a very common cause of bearing failure. With no room for expansion, friction heat develops in the lubricant and it deteriorates. If you are hand packing a bearing to be installed, never fill it more than half full of grease.

Ever wonder why some sealed type bearings have to be lubricated by a tube of grease instead of a grease gun fitting? One stroke of a grease gun can produce several tons of hydraulic pressure inside the bearing and force the seal or shield right out of its recess. Use the hand-squeezed tube of grease to prevent this type of damage to the bearing.

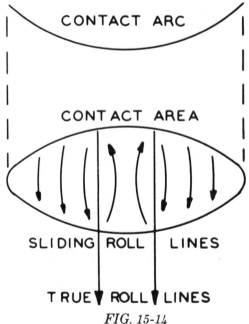

FIG. 15-14

MOTOR END PLAY

Sleeve bearing motors use metal and fiber washers to control end play. These can and do wear out and should be replaced if excess end play is noted. There is a definite reason for this, a reason that stems from the fact that an electric motor is a magnetic field device.

A magnetic field has a definite shape. The form of the coil windings, the size and shape of the soft iron core, and the placement of the windings in the core, contribute to the shape of the magnetic field.

The design of the rotor that is in the magnetic field determines the shape of the magnetic field that builds around the rotor from the EMF induced in it. Interaction between the two magnetic fields starts rotation and maintains rotation when a balance of opposed forces is reached. One of the results of a balance of forces is to center the rotor in the long axis of the magnetic fields.

This centering effect is not too strong and can be opposed by bearing friction, belt deflection, or even gravity. Thrust washers are installed on the rotor shaft to limit end play. EMF that is used to center a rotor in a magnetic field is wasted EMF. Thrust washers contribute to the efficiency of a motor. See that they are replaced when necessary.

ORDERING REPLACEMENT BEARINGS

Ordering replacement bearings means you must have all the identification numbers and letters from the old bearings. Different manufacturers use different identification systems. The possibility of replacing a bearing of one make by an identical bearing of another make can be determined by your bearing distributor if you have the available identification. As an example, let us take a radial Fafnir bearing with the identification M306KD. **M** indicates this bearing has a manufacturing tolerance of .0004 in., **3** means it is a single row, medium-duty type of bearing, **06** means the bore diameter is 30 millimeters, **K** means this is a Conrad type bearing, **D** means one side of the bearing has a grease shield.

If you have all the information, your bearing distributor can give you much better service in getting a replacement bearing. If you understand ball bearing construction and operation, you will be able to provide better service for your customers. The next time you are at your bearing distributor's, ask him for a copy of Fafnir Bulletin, Form 493, *How to Prevent Ball Bearing Failures, a Guide to Their Identification, Cause and Prevention.*

16

Commercial Refrigeration

Commercial refrigeration has been called the bread-and-butter work of the serviceman. It covers the servicing of the refrigeration units in grocery stores, drug stores, taverns, restaurants, warehouses, vending machines, and the like. It's hard work, sometimes dirty work, but dependable work. A good commercial serviceman will never lack for work.

I have heard quite a few servicemen say that they will not do commercial service work because they don't like to work on such dirty equipment, back in a corner behind boxes and cartons and with trash swept under and over the units. This is often the case, but I make it a point to clean up the equipment before I start work on it and when I leave the job. I charge the same hourly rate for this cleaning as I do for the rest of the work. To me, it is a necessary part of the service.

The basic reason too many servicemen do not like commercial service is that it requires the broadest range of knowledge of the basics of refrigeration, the widest range of application of these basics, the widest range of temperature applications both below and above freezing temperatures.

When the air conditioning system has ice or frost anywhere, it is a sign of trouble. A commercial system that operates at temperatures below 32 degrees produces ice and frost as part of its normal operation. The absence of frost on the suction service valve of the compressor may be a signal to the commercial serviceman that something is wrong. If you are operating with a 5 degree evaporator temperature and the TXV set for 10 degree superheat, suction gas temperature will be 15 degrees. This gas temperature has to produce frost from the moist air in the room. Commercial suction lines have to be insulated for this reason, and to prevent the system from picking up too much superheat and failing to cool the hermetic compressor.

Ice and frost also bring on mechanical problems when they build up in such places as the inside of a flare nut between the nut and the tubing. Repeated thawing and freezing will gradually push the tub-

SECTION OF FROST PULLED FLARE

FIG. 16-1

ing in until it pulls the flare out of its seat, Fig. 16-1. The result: loss of refrigerant and infiltration of air and moisture into the system in most cases. Because of the low suction temperatures, commercial systems are very susceptible to expansion valve freezeups and must be kept much drier than air conditioning systems.

The so-called frost-proof flare nuts have a shorter sleeve to support the tubing in back of the flare. Some long sleeve nuts have four holes drilled in them to allow moisture to drain out. The only real protection against freeze damage to the flare comes from insulation, insulating compounds and the like that prevent water vapor from contacting the cold metal. The serviceman must always make sure that he replaces or installs this seal after working on the system.

Ice buildup around an expansion valve can alter operation of the valve. Consider the fact that we are using the temperature of the remote bulb to build up a pressure in the refrigerant in the bulb. This pressure is transmitted through the capillary tube to exert force on the diaphragm of the valve. If the diaphragm is colder than the bulb, then it will take control of the valve away from the bulb. **Think about it.**

The effect on the TXV will be to throttle the refrigerant flow and reduce the coil temperature at the entrance to the coil. This causes lower temperatures and increases the ice buildup. It makes the problem worse. At the same time, superheat is increased and coil efficiency drops. If not corrected, ice will gradually build up over the entire coil and block the air flow.

It does not seem possible at first glance, but a small shortage of refrigerant in the system will also cause a coil to ice over. This is especially true on higher suction temperatures. Shortage of refrigerant means lower pressures. Lower pressures mean lower temperatures. Lower temperatures mean ice buildup that blocks air flow.

Less air flow means less heat input that means lower temperatures of refrigerant that causes more ice, and so on. It can lead to a vicious circle as you can see.

TXV's are designed for various temperature applications. They have operating springs that function best in the pressure ranges common to these temperatures. They have bulbs charged with types and amounts of refrigerant that do their work only in the desired temperature and pressure ranges. The commercial serviceman knows this and is careful to use only the correct TXV's for the particular system. He is careful to install the TXV's in the correct position. Some TXV's will not function correctly if installed on their side or with the diaphragm on the bottom. Some TXV's must be installed with an insulating and vapor-sealed cover over them. Some TXV's must not be installed in the coil area, but in a warmer space than the coil.

The commercial serviceman is not only familiar with the various types of TXV's, but he also knows the *how* and *why* of the application of this TXV by the equipment manufacturer to this particular piece of refrigeration equipment. He learns this by studying the service literature of the equipment manufacturer.

The air conditioning system, in almost all cases, is operating with a common evaporator pressure in all the coils. The commercial system may have a compressor that is connected to low sides (evaporators) with two or more distinct low side pressures. An ice cream cabinet at minus 10 degrees, a meat cabinet at 35 degrees and a produce cabinet at 45 degrees may all be connected to a single condensing unit, Fig. 16-2.

This is made possible by the installation of evaporator regulator valves (ERV) in the suction lines of the higher pressure low sides. ERV's allow refrigerant vapor to pass through them only down to a set pressure. The serviceman sets this pressure for the corresponding temperature at which the evaporator must operate.

If the room temperature is to be 45 degrees and the coil is designed to operate on a 10 degree TD, then the ERV pressure may be that which corresponds to 35 degrees for the particular refrigerant used. The condensing unit may operate on low pressure controls set at the desired temperatures for the lowest temperature low side.

This is not a common installation to run into. It has its problems. Oil return is one of them. Wide temperature fluctuations in the low temperature case can be another one. A high load in high temperature cases may bring the unit suction pressure above that needed for the low temperature case. Anti-return check valves in low tempera-

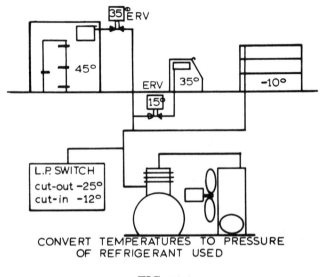

CONVERT TEMPERATURES TO PRESSURE
OF REFRIGERANT USED

FIG. 16-2

ture lines are not a satisfactory solution. The condensing unit must be oversized so that it will still operate at the lowest desired suction pressure with full cooling load on the high temperature cases.

This is not an economical operation. It is much better to group all low temperature cases on one condensing unit, medium temperature cases on another unit. The smaller temperature differences between cases can then be controlled with ERV's without requiring excessively large condensing units.

Control systems for this type of installation usually require solenoid valves and thermostats for each case, Fig. 16-3. The condensing unit may be controlled from a low pressure switch. Operation may be pumpdown control or one time pumpout. In some control systems, the condensing unit may be controlled by parallel relays operated by the case thermostats. As long as any one relay is closed, the condensing unit will run.

Some of the earlier air conditioning systems installed in homes and offices also used ERV's. These were console-type DX fan coil units such as the Frigidaire 3V or R50 and R100 units. Each fan coil unit had a thermostat and solenoid valve. A single ERV in the compressor suction line maintained evaporator pressure above freezing temperature. Compressor suction pressure varied widely depending upon how many of the DX console solenoid valves were open.

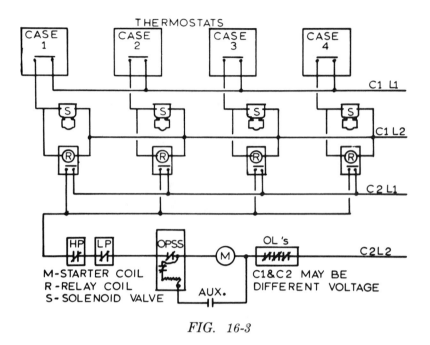

FIG. 16-3

Another special purpose valve found on commercial applications is the suction pressure regulator (SPR). Operation is identical with that of an acetylene or oxygen pressure regulator. The pressure on the **downstream** side of the valve cannot rise above the valve setting. Low temperature condensing units may not have motors large enough to operate at high suction pressures. SPR's throttle suction gas flow when needed to prevent overloading the motor at times when pressures may rise. This is usually right after a defrost cycle or when a case is being loaded with fresh supplies of produce, milk, beer, or what have you. The SPR spreads the load out over a longer pulldown time period.

DEFROSTING

Commercial refrigeration systems must maintain temperatures that require a refrigerant temperature in the coil that is well below freezing. To maintain a room or case at 40 degrees means we must supply air from the coil that is below 40 degrees. Say we supply 30 degree air. To supply 30 degree air means the refrigerant temperature must be below 30 degrees. If the coil was selected on a 10 degree TD, then we must have a 20 degree refrigerant temperature. A 15 degree TD means a 15 degree refrigerant. All of which means that we are going to collect ice or frost on the coil.

Coil temperatures close to freezing mean that moisture vapor will be condensed to a liquid and then frozen in a clear film of ice on the surface of the tubes (prime surface) and fins (secondary surface). Ice is a fairly good conductor of heat energy. Ice builds in thickness slowly and melts slowly.

Coil temperatures well below freezing will freeze moisture vapor before it has a chance to condense to a liquid. This results in frost instead of ice. This white frost is so porous that it acts as a very good insulator around the tubes and fins. The decrease in heat energy transmission to the refrigerant lowers coil temperature and helps to create more frost. A thick coating of frost blocks air passage and this cuts down further on heat transmission. Soon the coil is a block of frost with a film of ice forming over the outer surface to coat it solid.

All of this means that the defrost cycle is an important and integral part of most commercial refrigeration systems. Let's take a look at some of the defrost cycles and defrost systems.

The simplest defrost cycle of all is where the low pressure switch is set to cut the compressor off at a pressure that insures the box or case will be down to the desired temperature. The cut in point is at a pressure that cannot be reached as long as there is ice or frost on the coil. Remember that the temperature of the coil determines the pressure of the refrigerant in the coil and pressure is transmitted in all directions in a fluid.

This is a simple control system and it will only work with a simple installation: one case and one compressor or two cases of equal size and loading, operating at the same temperature, with one compressor. Every cycle of the condensing unit is a defrost cycle.

When you get away from simple installations you are likely to find time clock controls and defrost thermostats. Time clocks are set to turn off the fans and sometimes the compressors for a set time period once or twice every 24 hours. On multiple lowside installations, the time clocks may turn off individual lowsides at staggered time intervals. They may also close liquid line solenoids and stop or start coil fans. As a general rule, they stop the fans to keep from warming the products stored in the box or case.

Defrost thermostats have two sensing elements. One element senses the air returning to the coil fan and is set for the desired unit temperature. The other element is usually placed between the fins (secondary surface) at a location where it will be covered with ice and frost as they build up on the fins. When this element is covered up it stops the compressor or opens the coil solenoid, and will not reset until the element is warmed to a temperature above freezing. The principle of the defrost cycle has been placed in a small package.

Some evaporators have electric heating elements for defrosting. These heating elements may be mounted below the coil or may even be inserted inside the bottom tubes of the coil. The time clock stops the fan, closes the solenoid, and turns on the heaters, Fig. 16-4. A thermostat may or may not be used to terminate the defrost cycle. If a thermostat is not used, the length of the defrost cycle must be timed on the basis of experience. The cycle must be long enough to insure complete defrosting and not too long or the case or box will warm up too much.

Electric defrost can be used on installations where the temperature of the space is held below freezing. In these installations, you will find that the drain lines and drain pans are also heated. On larger commercial installations, you will find that the coil housings are insulated and that units are equipped with motor operated, insulated doors. These doors are closed during the defrost cycle to keep from losing too much heat to the cold room. Just one more item to be checked and serviced with this type of defrost unit.

All of these defrost methods boil refrigerant out of the evaporator during the defrost cycle. It would seem that a solenoid valve in the suction line would build up pressure and keep heat energy in the coil for a quicker defrost. The theory is true. **But,** when you open the suction solenoid, you are going to have a coil full of liquid refrigerant that will slug the compressor and wreck it in short order. This is also true of any method of hot gas defrosting. In hot gas defrosting, a so-

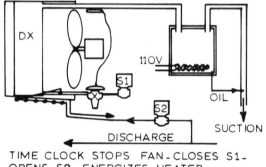

TIME CLOCK STOPS FAN-CLOSES S1-
OPENS S2-ENERGIZES HEATER.

FIG. 16-4

lenoid valve in the hot gas line opens to admit hot gas from the compressor discharge into the line between the TXV and the coil. On low temperature installations, the hot gas line heats the drain line and the drain pan before it enters the coil.

The hot gas will lose pressure and condense to a liquid due to the low temperature refrigerant in the coil. As this refrigerant warms up, the pressure will rise. This will force the liquid out of the coil and back to the compressor.

Any defrost method that forces liquid refrigerant out of the evaporator must have a suction accumulator to trap this liquid and prevent slugging the compressor. It should also have some means of slowly vaporizing the liquid out of the accumulator and getting it back into the refrigerant cycle, as in Fig. 16-4.

Some cases have an external blower coil that serves as an accumulator and uses storeroom heat to vaporize the liquid. Others have an accumulator with an electric heating element in it. The serviceman has to check the operation of all parts of a system like this. The failure of a heating element or the element thermostat can throw the whole system out of balance.

Larger commercial coils, including ammonia coils, may use a system of hot gas defrost that includes two pressure-operated valves. These are usually flooded coils that have a suction surge drum on the

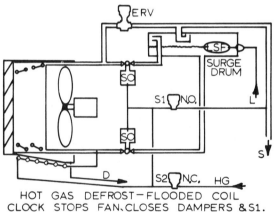

HOT GAS DEFROST—FLOODED COIL
CLOCK STOPS FAN. CLOSES DAMPERS & S1.
OPENS S2. HG PRESSURE CLOSES SO
VALVES. HG HEAT WARMS DRAIN LINE,
PAN, LIQUID IN COIL, ERV PASSES VAPOR
ABOVE FREEZING PRESSURE TO SUCTION.

FIG. 16-5

coil and a low side float feeding the drum. You will always find that the coil is connected to the drum with a bypass line with an ERV in it. The ERV is set to operate at a pressure above the freezing point. Study the drawing, Fig. 16-5. You will see that it is a logical operation that obeys the laws of gases. Hot gas pressure closes the valves. Suction pressure opens the valves. The coil remains flooded during the defrost cycle and is ready to resume cooling very quickly after the cycle ends. Those that I have worked on were manufactured by Refrigeration Specialties Co. of Chicago.

Of all the hot gas defrost systems I have worked on, my favorite has been the Kramer Trenton *Thermobank*, Fig. 16-6. The *Thermobank* is a suction accumulator inside of a larger tank. The outer tank contains a glycol solution and a hot gas coil. During normal operation the hot gas from the compressor goes through the glycol tank line and heats the brine to the hot gas discharge temperature. The hot gas then goes to the condenser for normal condensing of the refrigerant.

When the time clock starts the defrost cycle, the evaporator fan stops and the hot gas solenoid opens. Liquid is forced from the coil to the *Thermobank* accumulator. The stored heat in the brine boils off the liquid and maintains a fairly high suction pressure to keep the hot gas temperature high. Defrosting is very fast. Defrost times vary from 3 to 6 minutes and I have actually operated some units with a 1.5 minute defrost every 6 hours and done a good job.

Properly installed and timed, the *Thermobank* is very dependable and economical to operate. It is not foolproof, however. I was once

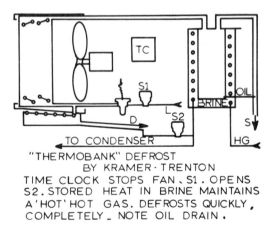

"THERMOBANK" DEFROST
BY KRAMER · TRENTON
TIME CLOCK STOPS FAN . S1 . OPENS
S2 . STORED HEAT IN BRINE MAINTAINS
A 'HOT' HOT GAS. DEFROSTS QUICKLY,
COMPLETELY _ NOTE OIL DRAIN .

FIG. 16-6

called in on a locker plant job I had designed but not installed due to a lower bid.

The low bidder had installed the three blower coils and the 10-hp condensing unit, but he had tried to cut corners by only using one of the coil *Thermobanks*. Each *Thermobank* is designed to match its companion coil. The one *Thermobank* would not store enough heat, would not hold enough refrigerant liquid, and would not pass enough hot gas through the small line to do any defrosting. I quoted a price on repiping the job and adding the two missing *Thermobanks*, but the customer went bankrupt before he could get it done.

Regardless of the type of defrost system, a poor installation, or failure to check out all the parts of the system, can cause it to fail. I was part of a crew called in to remove ice from the floor of a frozen food warehouse one time. The defrost system was a water spray, Fig. 16-7. The time clock shut off the fans, closed the doors, and opened the solenoid valve in the water spray line. Everything worked fine except that the electric heat cable around the drain line had burned out. The water overflowed the drain pan and made a skating rink out of the floor. It took eight men with scrapers and welding torches and shovels to get a dry floor again. The room rose from minus 20 degrees

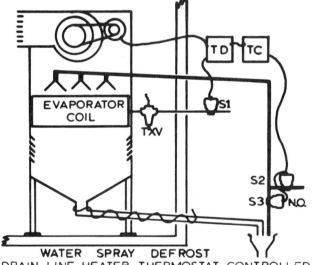

WATER SPRAY DEFROST
DRAIN LINE HEATER THERMOSTAT CONTROLLED.
TIME CLOCK STOPS FAN. CLOSES S1- S3.
OPENS S2. TD RELAY ALLOWS DRAIN
PERIOD BEFORE REFRIGERATION STARTS

FIG. 16-7

to plus 26 degrees before we got the job done. Fortunately, we were able to move all the food to another room first. Always check the operation of these heating cables with an ammeter to make sure they are working.

CONDENSING REFRIGERANT

The commercial serviceman is going to run into a wide variety of condensing methods. Condensing mediums are still either water, air or a combination of the two. A small condensing unit with a water-cooled condenser is not uncommon in a small grocery store of the *ma-and-pa* type. The water is used in one pass through the condenser and then wasted to the drain.

How much water do they use? Remember that one Btu is the amount of heat necessary to raise the temperature of one pound of water one degree Fahrenheit. If the water comes in at 60 degrees and leaves at 95 degrees, each pound of water absorbs 35 Btu. There are approximately 8 lbs. of water to a gallon. (A pint's a pound the world around.) A gallon of water, then, would carry 8 x 35 or 280 Btu. 12,-000 Btu divided by 280 means almost 43 gallons of water are required to condense a ton of refrigeration.

The commercial serviceman never makes the mistake of calling a 1-hp condensing unit a one ton unit. Tonnage depends on the weight of refrigerant pumped. The *weight* of refrigerant pumped by a compressor depends on the density of the gas. At air conditioning suction and head pressures, it takes approximately 1-hp to pump the weight of refrigerant necessary to carry 12,000 Btu. At the lower suction pressures necessary for commercial refrigeration, the same cubic displacement compressor, pumping a gas half as dense, would only pump enough refrigerant by weight to carry 6,000 Btu. If the suction pressure was always kept at this low reading, you could power the compressor with a half-horse motor. Commercial condensing units are available in low, medium, and high back pressure ratings.

As an example of the effect of low density gas on tonnage, I will mention one I worked on a good many years ago. First stage: two 8-cylinder compressors with 125-hp motors driving each. Second stage: one 6-cylinder compressor with a 75-hp motor. Refrigerant was R-12, box temperature minus 105 degrees F, total horsepower 325. Refrigerating effect was nine (9) tons.

Condensers using city water accumulate scale at a slow rate. Since they are usually double tube condensers, or shell and coil condensers, they have to be acid cleaned to remove scale. The water valves, controlled by head pressure or temperature, require cleaning of the inlet screens and polishing of the internal operating piston almost every year. Most always these water valves are set to maintain a 105 degree condensing temperature of the refrigerant.

Air-cooled condensers, of course, must be kept clean—not only on the surface, but deep between the fins. Condenser fan blades must be clean and the fan motor will burn up if dust and lint accumulate to block cooling air flow through the motor. Surface dirt can be brushed off. Internal dirt can be blown out with a refrigerant or nitrogen blast.

Combination air and water-cooled condensers are not unusual on commercial condensing units. Since the water-cooled condenser only comes into use when the ambient air temperature is high, the water valve on these units usually gives a lot of trouble. They tend to stick from not being used. If you are checking the unit during cool weather, block the air flow over the air-cooled condenser to bring the water valve into operation. This is a quick and simple test to make. Of course, you must have a high pressure gauge attached at the time.

The most important test to be made with any water valve is the check to make sure it closes off normally. When the grocer suddenly gets a water bill that is about five times higher than normal, don't think you won't hear about it.

If your hearing is good, you can tell when a water valve shuts off by listening with a screwdriver linking your ear to the valve. I use an amplifier made from an old-style telephone ear piece with a length of brass brazing rod riveted to the metal diaphragm. Army surplus stethoscopes are fine for this purpose, too. If the valve does not shut off properly, take it apart and clean and polish it. Repair parts, seats, and washers are available at your wholesaler. You can always justify the repair charge for a water valve by asking if they would rather pay the high bill for wasted water, plus the sewer tax on the water bill.

Commercial refrigeration systems use cooling towers in some installations with all of the cleaning, algae, and scaling problems that come with any tower installation. In addition, you have a problem that is not too common on air conditioning installations: winter operation. This requires an indoor sump. The tower operates with what is called a dry basin, Fig. 16-8. You sometimes find tower lines with electric heating cables under insulation that must be checked for correct operation. Cables do not go bad as often as buried thermostatic switches do. Thermostat-controlled tower fans are the rule here, as are thermostat-controlled valves that bypass warm water from the top deck of the tower and direct it across the basin only for cooling.

Evaporative condensers are common on commercial installations. They may be indoors, in which case you have problems with air supply and discharge. They may be outdoors, in which case you have the dry sump, indoor sump problem. They may be operated as air-cooled condensers in low temperature periods. This brings on the problem of thermostat-controlled dump valves and water line shutoff and drain valves. Thermostat-controlled fans also bring on excessive tube scale problems.

One complaint I have heard recently is brought on when someone sees the steam cloud from a tower or evaporative condenser and hollers, "Air pollution." The truth of the matter is just the opposite. That cloud is composed of pure water vapor evaporated from the heat rejection device. You are actually removing pollution from the air. All the air that passed through the tower or evaporative condenser has been washed and cleaned of the biggest part of the carbon, sulfur, and nitrogen oxides; also, soot, dust and other particulate matter.

Evaporative condensers and cooling towers are very effective tools in fighting air pollution. The serviceman, however, has his problems with condenser pollution.

Air-cooled condensers are widely used in commercial refrigera-

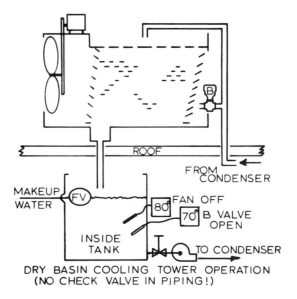

DRY BASIN COOLING TOWER OPERATION
(NO CHECK VALVE IN PIPING!)

FIG. 16-8

tion. Many of the newer installations use multicircuit condensers. A single air-cooled condenser may have as many as 20 separate condensing circuits, one circuit for each condensing unit.

Winter operation of multicircuit air-cooled condensers brings the usual subcooling headaches that cause coil freezeup and poor TXV operation because of loss of pressure differential across the TXV. The problems of maintaining head pressure on a single circuit condenser are bad enough. Controls may reduce condenser capacity by flooding the condenser tubes with liquid, cycling fans, or restricting air flow with dampers. When you have a multicircuit condenser, these problems are compounded by the differing needs of each circuit being served by a single condenser. The serviceman must really rack his brain to come up with a suitable compromise that will best serve all the compressor circuits.

A recent development in supermarket refrigeration is helping to alleviate this problem. The air-cooled condensers are being used to supply winter heating to the building. It is a logical use of this wasted energy, and the savings in the cost of heating fuel will usually pay for the extra expense of the ductwork necessary to convert from outside air, through the condenser, in summer to recirculated room air, through the condenser, in winter. Getting away from the extreme low winter air temperatures through the condensers does away with practically all of the condenser pressure problems. It does mean extra maintenance work in that condensers must be kept cleaner, must have air filters for recirculated air, and must have fans capable of moving a sufficient volume of air at ductwork pressures. Needless to say, engineering design of these installations is critical.

REFRIGERATION AND SANITATION

No discussion of service problems would be complete without getting into sanitation. It is one of the main reasons for refrigeration and we should know *why* and *how* of the subject if we are to think logically about service problems that arise in this area.

I have heard servicemen make the statement that there are some foods that refrigeration will not protect. They are correct in making this statement, but I wonder if they know why they are correct. A knowledge of the process of food decay is necessary to understand how refrigeration protects food. This knowledge will help the serviceman materially in his work.

Microbes, microscopic organisms found everywhere in nature, cause food decay. Some microbes are plants and some are animals. Under the right conditions, when moisture, warmth, and sometimes oxygen are available, microbes can grow and multiply enormously. Three groups of microbes are principally responsible for food decay. They are yeasts, molds, and bacteria. One bacterium present in meats, cream sauces, milk or liquid food leftovers can multiply to 2 billion in 24 hours providing the food is warm enough for the bacteria to be mobile and move among the food cells. Refrigeration reduces this mobility and thus slows down food decay.

Animal bacteria digest food as they grow and multiply. As they digest food they eliminate waste products. These waste products are called toxins. Toxins cause food poisoning. Once toxins are present in food they cannot be removed by heating or by refrigeration of the food.

Remember these facts: heating (pasteurization or cooking) kills bacteria but does not remove toxins; refrigeration effectively slows down food decay but does not remove toxins; neither heating nor refrigeration will prevent food from acquiring new bacteria if it is exposed to them.

Microbes can live for years under favorable conditions. Under unfavorable conditions, some microbes die and others dry up and become spores. These spores can travel through the air and if they manage to come to rest on a favorable surface will become active, grow and multiply.

Not all microbes are harmful. Yeasts are necessary for the production of beers and wines and in the making of breads. Uncontrolled yeasts can also spoil foods. Molds are necessary for the production of some cheeses and medicines. Of course, we are familiar with the spoiling of foods by other molds. Beef is *aged* or tenderized by beneficial bacterial action under controlled temperatures.

Refrigeration is one of our most widely used methods of preserving foods, but we can see from this knowledge of food decay that refrigeration must be accompanied by good sanitation practices for it to do the most good. Refrigerators must be clean. When spilled foods accumulate on interior surfaces they may become obvious molds that one would notice and wipe off. They might also dry up and microbes in the food become spores that float through the air to reinfect fresh foods as they are placed in the refrigerator.

Other employees may be responsible for case, box or room sanitation and cleaning, but the serviceman is still the one who knows the *how* and *why* of cleaning and sanitation of the refrigeration evaporators and fans. All too often, the blower coils and fans are enclosed in such a way that the serviceman is the only one qualified to open them up for cleaning, cleaning them, and then reinstalling the covering panels. Cleaning and sanitizing drain pans and drain lines is particularly important in food processing and storage. A drain pan opening on the *pull* side of the fan can pull bacteria through the line from the other end unless it is properly trapped and connected. **Think about it.** A properly designed trap has dimensions that hold a height of water greater than the fan suction measured in inches of water by the serviceman's manometer.

When the serviceman gets a complaint that food spoils in the refrigerator and temperatures are within the normal range, he should look to lack of sanitation as a possible cause. A visual inspection may be all that is necessary although a few good sniffs is enough if you have a sensitive nose. The remedies are removal of any containers of spoiled food tucked away in the back and thorough cleaning of the interior surfaces. Soap and water washing will remove greasy substances and dirt. Some like to use a vinegar and water rinse for odor removal. This can be done but you must remember that when vinegar dries it leaves vinegar spores that can change some foods. I prefer to rinse with a chlorine bleach solution and then follow with the old standby, a baking soda solution rinse.

One thing for certain, if you do find this condition in a refrigerator, you are going to be in a spot. It will take a lot of tact on your part to tell a housewife what you think is causing the food spoilage. I generally take the attitude that since everything else has been checked out and found to be correct, the possibility remains that these highly invisible microbes have somehow gotten into the refrigerator and the only way we can eliminate this possibility is to sanitaize and see if this does the job.

The commercial refrigeration serviceman is much more familiar with the need for sanitation in coolers than those doing domestic service. Many of us will not eat food, especially meats, served in strange restaurants and even on some airline flights because the smell was the same as the one found in these coolers. When these coolers are cleaned and sanitized, the work is usually done by someone other than the serviceman. One part they will miss is the cooling coils. Fin coils and fan blades in these coolers will collect fats, dust, and microbe spores on their surfaces. The serviceman should point this out to the owners and make it clear to them that he is the one who is equipped and trained to clean these units.

We read about the development of fungicide sprays for air conditioning systems because some people are allergic to fungus and mold that grow on the cooling coils. With all due respect to the doctors, I would like to point out that these are products of microbes present in the air and they must have a surface on which to light that is not hostile to their growth. In other words, they will not grow and multiply unless they land on a dirty cooling coil or a partially flooded drain pan.

Before I recommend using such a product to a customer, I am going to make sure that the coil, the drain pan and the duct work is clean. If it is not, then I am going to ask that this cleaning be done. Sanitation must come before any other treatment. Spraying with coil cleaning detergents, rinses with clear water, chlorine water, and soda water will help to clean and sanitize these surfaces so that mold and fungus spores do not have favorable breeding places. Then a fungicide will have a chance to do the job. **Think about it.**

SUCTION ACCUMULATORS

The commercial serviceman will run into a wide variety of special purpose devices made necessary by the nature of the application. A suction accumulator or surge drum is a special purpose device that traps liquid refrigerant and prevents it from slugging the compressor. A suction **gas** accumulator is another device that is used to prevent a condensing unit from cycling too often in some cases. It was originally developed for use on instantaneous DX water coolers. Everytime you draw a glass of water from the cooler, you boil off a small amount of refrigerant. Without a gas accumulator, you would have a very short on cycle of the compressor. With the gas accumulator installed, a lot of water must be drawn to build up enough pressure in the drum to cycle the compressor. You get longer *on cycles*, at wider spaced intervals, from the condensing unit. The empty space to be filled in the drum still furnishes suction *pull* with the compressor off.

ICE MAKERS

Present day package ice makers require a wide knowledge of refrigeration, water pumping, solids concentration in water, water sanitation, and mechanical ice scraping, crushing, and collecting, not to mention defrosting, gear reducers, and sharpening of scrapers and cutter bars.

Making artifical ice is one of the oldest uses of refrigeration and properly belongs in the industrial refrigeration field. The making of *crystal clear* ice with cans in brine tanks, air bubbler lines, core suckers and all the rest of the equipment is worth a book in itself. The use of the packaged ice makers has reduced the ice cake business considerably. The need for a knowledge of ice making has been transferred to the commercial serviceman. The fundamentals of ice making remain the same.

When you evaporate water, the vapor is pure water. The solids remain behind in the liquid and build up in strength. This is reason for a constant bleed in evaporative cooling, condensers and cooling towers. Steam boilers require routine blowdown for this same reason.

When you freeze water, the same thing happens. Salts and other solids lower the freezing point of water. Pure water freezes first. As the pure water freezes, the solids are then concentrated in the remaining liquid. The opaque core found in the center of a cake of ice was the water that froze last, and at a lower temperature, because it contained the concentrated solids. The frequency with which the *core sucker* was inserted in the ice cans, to remove the concentrated water, determined the size of the core.

Today's ice makers pump water over a plate or drum, or through a tube. The ice freezes in a film that gradually grows thicker. This is actually a better way to make clear, pure ice than was possible in the cans. The solids are still concentrated in the liquid and must be bled off and replaced with fresh water. If this is not done, you will eventually wind up with opaque ice that contains solids. Every ice maker contains a bleed off device of some kind. It may be a simple measured flow tube to a drain. It may be an overflow tube in the sump. It may be a solenoid valve. The serviceman must understand that is it a necessary part of the operation. He must understand how it works and how to check it for correct operation.

The operating suction pressure of the refrigeration cycle is also important in the making of clear ice. Freezing at too low a temperature will prevent removal of the concentrated solids. Refrigerant charge is usually critical in these small systems and ERV's are quite often installed in the refrigeration system.

Defrosting is a necessary part of the ice making procedure. Ice will not slide off a plate or slide down a tube unless the metal is warmed to provide a film of water between the ice and the metal. This is the lubricating film that allows gravity to move the ice. The rotary freezing drum of the York Flakeice Machine does not defrost to remove the ice. The flakes of ice are scraped off by a sharp cutter bar properly positioned. The multiple refrigerant shaft seals on this rotating drum can provide a good workout for a serviceman when they start leaking.

Sanitation plays a very important part in ice making. It is surprising how many forms of algae and slime can grow in cold water. The water tanks, pumps and water lines must be kept clean and sweet. Water that stinks makes ice that stinks. When the glass of drinking water smells sour or *woody*, you can bet that the ice maker water sump is plenty dirty.

Failure of the constant bleed or insufficient constant bleed will cause lime scale to form on the freezing surfaces even at low temperatures. This scale will cause ice sheets or rods to *hang up* and fail to move off or out on the defrost cycle. Scraping the scale off is out of the question. The surfaces must be smooth and well polished for ice to slide off.

Special slimicides and scale removal acids are made for ice makers. Vinegar is an acid that will remove scale. Citric acid will remove scale. Both of these are approved for use in ice makers in the proper limited concentrations. Use the proper, approved chemicals for slime and scale control in ice makers in the limited quantities recommended. Keep the equipment as clean as possible. Sell the owner on a regular maintenance checkup. It pays off for both the serviceman and the customer.

The commerical serviceman will find that he covers the whole wide range of refrigeration applications. He will find a great variety of mechancial equipment linked to the refrigeration equipment. He will find that he needs to apply a great deal of basic knowledge of physics, chemistry and eletricity and mechanics in solving his problems. And he will still find that the biggest job of all in commercial service will be that of cleaning the equipment and keeping it clean-just as in any other air conditioning, heating and refrigeration application.

Afterword

When a serviceman arrives to check out a refrigeration or an air conditioning system, he carries his tool kit, a gauge manifold and a few thermometers in his shirt pocket. There is a malfunction within the system, and he is there to find out what it is.

The serviceman will use his gauges to read the pressures in the high and low sides of the system. He'll then use his thermometers to read temperatures at various points in the system.

The information the serviceman gets from his instruments will register in his mind, which already contains a fundamental working knowledge of just how the system ought to work, and his grey matter will churn out a solution to the problem at hand.

The obvious, often overlooked details can tell you, the serviceman, so much. I remember one time when I checked temperatures on a Carrier centrifugal that was not doing the job it should. Pocket thermometers I'd placed in the holes of the casting indicated the liquid refrigerant coming out of the economizer was far too warm. On the basis of the temperatures I read, I trusted my instincts and pulled all the R-113 out of the system and removed the plate over the economizer and high side float valves. Consequently, I found about a cupful of steel shavings trapped under the economizer float. The shavings were holding the valve open. The stuck valve sure made a difference in the chilled water temperature.

The lesson? Aided by common instruments such as pressure gauges and thermometers, simple observation is essential in gaining the information that will enable you to diagnose problems with any refrigeration or air conditioning system. All you have to do as a troubleshooter is retrieve basic information, in-put that data into the complex computer of yours called the mind, and **think** about it!

There will never be a substitute for the serviceman in the air conditioning, heating and refrigeration field. No computer ever built can approach the total complexity of the human mind. Powered by the basic electricity derived from electrons in motion, the brain has capabilities far beyond any man-made machine. Self-programmed by a process we call thinking, human grey matter hosts an almost limitless memory bank for storing information. The brain also offers random access to the most miniscule bits of stored data, and it can instantaneously retrieve information from countless books, service manuals and catalogues. No, I don't think there will ever be a substitute for the **thinking serviceman.**

Index

Air, combustion, 112
Air cooled condenser, 6
Algae, 87
Ammeter, 135
Anomometer, 107
Atmospheric tower, 80
Automatic expansion valve, 59
Bacteria, 244
Ball bearings, 221
Bearing failures, 225
Bimetallic switch, 161
Brazed tube condenser, 124
Brazing, 195, 203
Buck-boost transformer, 181
Burnouts, 215
Capacitors, 145
Capillary tube, 59
Charles' Law, 16
Circuit breakers, 158
Circuits, buck-boost, 181
Circuits and controls, 147, 151, 173
Circuits, transformers, 151
Circulation, air, 93, 95
Coil freezing, 5
Combustion air, 112
Condensers, 2, 115
Air cooled, 61
Cleaning, air cooled, 35
Cleaning, water cooled, 77, 127, 130
Evaporative, 85
Pressure loss, 31
Purging, 38

Repairing, 124
Retubing, 116
Scale, 70, 76, 84
Conduction, 16
Connectors, 148
Conrad bearing, 222
Constant bleed, 79
Contacts, switch, 148
welded, 168
Cooling towers, 4
Atmospheric, 80
Constant bleed, 79
Dry basin, 243
Evaporation, 78
Forced draft, 83
Dalton's Law, 35
Defrosting, 235
Defrosting cycle, 236
Desuperheating, 33
Distilling refrigerant, 64
Driers, 63, 190
Electrolytes, 73
Electronic treaters, 90
End play, 229
Evaporation, 78
Evaporative condenser, 85
Evaporator regulator valve, 233
Float valves, 60
Forced draft tower, 83
Gas laws, Charles', 16
Boyle's, 18
Dalton's, 35
Hardness, water, 68
High side float, 60

Hot gas defrost, 238
Humidity, 44
Ice makers, 248
Instruments, 133
Klixon, 160
Leak detection, 205
Leak repairing, 201
Lifetime bearings, 220
Lockout relay, 169
Manometer, 102
Megohmmeter, megger, 137
Meters, 133
Microbes, 245
Multimeter, 133
Oil pressure safety switch, 176
Oil separator, 212
Oil traps, 210
One time pumpout, 164
Overload relay, 159
Oxidation potentials, 71
Pilot solenoid, 179
Psychrometrics, 41
Radiation, 15
Refrigerant controls, 56
Refrigerant tables, 22
Reheat, 49
Ring oiler, 219
Relative humidity, 44
Relay, lockout, 169
  overload, 159
Resistance thermometer, 134
Retubing condensers, 116
Sanitation, 244
Saturated state, 19
Scale, condenser, 70, 76, 84
Sensing tube, 108
Sleeve bearings, 217
Softeners, water, 91
Solenoids, 178
Specific Humidity, 44

Spray nozzles, 81
Subcooling, 30
Suction pressure regulator, 235
Suction pressure accumulator, 247
Superheat, 28
Switches, 149
Bimetallic, 161
Oil pressure, 176
Solenoid, 180
Warp, 160
Tables
Oxidation potentials, 71
Pressure conversion, 104
Refrigerants, 22
Relative humidity, 44
Water, 20
Thermobank, 239
Thermodynamics, 14
Thermostatic expansion valve, 56
Thermometer, resistance, 134
Three way switch, 162
Three wire control, 163
Transformer, connections, 151
  buck-boost, 181
Treaters, electronic, 90
Treatment, water, 89
Tube expander, 121
Tube, sensing, 108
Tubing, copper, 197
Unloaders, 2
Vacuum Pumps, 187
Vapor dome, 65
Velocity meter, 108
Water, 67
Water spray defrost, 240
Watt meter, 240
Warp switch, 160
Wick oiler, 219

For more valuable HVAC/R information...

The Schematic Wiring Book Set

Electrical diagrams are easy to read, if you know the graphic symbols, the language of the legend and where to start. This self-paced workbook set will help you learn to understand even the most complicated wiring diagrams and teach you what you need to know before you make another service call. And once you've learned electrical shorthand, **you'll be able to read any schematic.**

These step-by-step guides include sections on the use of circuit-testing instrument: the voltmeter, ammeter and ohmmeter, with easy to understand instructions on how to use them effectively so you can troubleshoot electrical malfunctions quickly and accurately.

With the aid of this set, you'll learn by actually creating and drawing schematics, defining the functions of equipment involved, listing the electrical components and tracking the electrical sequence of systems. Practical simulations will challenge you to find the source of common system failures and offer you the chance to sharpen your analytical skills by comparing your hunches with the actual solutions.

Included in this two book set are: **Schematic Wiring Simplified,** a workbook that teaches electrical shorthand, components and their functions, sequence of operation and the intent of circuitry, and **Understanding Schematic Wiring** a complete guide that details the electrical installation of residential heating and air conditioning units.

Take this opportunity to master the language of wiring and learn how to **troubleshoot electrical diagrams like an expert.**

HVAC/R Reference Notebook Set

Here's an indispensable reference set for contractors, servicemen or newcomers to the heating-cooling industry who don't want to lug bulky reference texts around from job to job. This convenient wallet-sized reference set will give you information whenever and wherever you need it.

The set includes a comprehensive 128-page **ACHR Dictionary,** a 176-page **A/C, Heating Reference Notebook** and a 160-page **Refrigeration Reference Notebook** – both notebooks complete with all the tables and charts you refer to daily.

Don't get caught short on your next job because you don't have a convenient **reference library in your hip pocket!**